Vadim Salnikov

Paramural bodies and other entities of the plasma membrane

AF167397

Vadim Salnikov

Paramural bodies and other entities of the plasma membrane

Paramural bodies and other entities of the plasma membrane at the formation of the secondary cell walls of plants

LAP LAMBERT Academic Publishing

Impressum / Imprint

Bibliografische Information der Deutschen Nationalbibliothek: Die Deutsche Nationalbibliothek verzeichnet diese Publikation in der Deutschen Nationalbibliografie; detaillierte bibliografische Daten sind im Internet über http://dnb.d-nb.de abrufbar.

Alle in diesem Buch genannten Marken und Produktnamen unterliegen warenzeichen-, marken- oder patentrechtlichem Schutz bzw. sind Warenzeichen oder eingetragene Warenzeichen der jeweiligen Inhaber. Die Wiedergabe von Marken, Produktnamen, Gebrauchsnamen, Handelsnamen, Warenbezeichnungen u.s.w. in diesem Werk berechtigt auch ohne besondere Kennzeichnung nicht zu der Annahme, dass solche Namen im Sinne der Warenzeichen- und Markenschutzgesetzgebung als frei zu betrachten wären und daher von jedermann benutzt werden dürften.

Bibliographic information published by the Deutsche Nationalbibliothek: The Deutsche Nationalbibliothek lists this publication in the Deutsche Nationalbibliografie; detailed bibliographic data are available in the Internet at http://dnb.d-nb.de.

Any brand names and product names mentioned in this book are subject to trademark, brand or patent protection and are trademarks or registered trademarks of their respective holders. The use of brand names, product names, common names, trade names, product descriptions etc. even without a particular marking in this work is in no way to be construed to mean that such names may be regarded as unrestricted in respect of trademark and brand protection legislation and could thus be used by anyone.

Coverbild / Cover image: www.ingimage.com

Verlag / Publisher:
LAP LAMBERT Academic Publishing
ist ein Imprint der / is a trademark of
OmniScriptum GmbH & Co. KG
Heinrich-Böcking-Str. 6-8, 66121 Saarbrücken, Deutschland / Germany
Email: info@lap-publishing.com

Herstellung: siehe letzte Seite /
Printed at: see last page
ISBN: 978-3-659-76451-6

To Dr. Andrei Vasilyev,
the great scientist of cytology,
dedicated...

Salnikov V.V.

Paramural bodies and other entities of the plasma membrane at the formation of the secondary cell walls of plants.

ATLAS - MONOGRAPH

2015

The Ministry of education and science of Russian Federation

Kazan (Volga region) Federal University, Kazan University, KFU

Kazan Institute of Biochemistry and Biophysics Kazan Scientific Centre of the Russian Academy of Sciences

Content

Author's foreword

Since we began the study of cell wall formation over 30 years ago, particularly the cell wall ultrastructure, we have often observed in micrographs what we believe to be plasmatubules between the plasma membrane and cell wall. The data we have accumulated has raised questions concerning the role of these paramural bodies in the plant cell. When we first reviewed the literature on the subject, we could not find a definitive answer on this matter, to our surprise. We then came across a micrograph taken with a confocal microscope (Fig. 23 - Photograph courtesy of Brian Gunning. Plant, Cell & Environment Volume 25, Issue 2, pages 239-250, 11 MAR 2002 DOI: 10.1046 / j .0016-8025.2001.00808. x), showing a pattern of intense plasmolysis in onion scale fixed cells - a classic object for the study of conventional plasmolysis. Used to observing the more 'modest', classic, light microscope images of plasmolysis where cell plasma membranes are seen with convex or concave cytoplasms, we were quite surprised to observe in the confocal micrograph the clearly visible paramural bodies that stretched between the plasma membrane and the cell wall. The question immediately arose as to how such a restructure of the plasma membrane could be possible in terms of ultrastructure physiology, and how the formation of such a large number of Hechtian strands could occur.

After observing hundreds of pictures (by electron microscopy in works since 1975) of different plant fixed for the study of their ultrastructures that were available in laboratories, we noticed a similar effect in similar structures. Thus we called these structures between the plasma membrane and cell wall plasmatubules.

Based on our own data and that in the literature, we have attempted in this publication to analyze existing issues and present our understanding regarding some of the questions that remian unclear, for example, regarding the terminology - how certain structures between the cell wall and plasma membrane should be properly named, how plasmatubules, Hechtian strands and other paramural bodies develop in a plant cell, and what functional load they carry.

Keywords: paramural bodies, plasmalemmasomes, plasmatubules, vacuolar tubules, Hechtian strands, transfer cells, plasmolysis, compensatory endocytosis, the secondary cell wall, primary phloem of the bast fibers, tension wood of poplar, hairs of the cotton seed, secondary phloem of the bast fiber of hemp.

Introduction

Structures within the cell membrane that are localized between the plasma membrane and cell wall are called paramural bodies. This term in the Russian language translation implies near-wall structures or entities between any two party-limiting barriers. Paramural bodies are divided into two classes on the basis of their origin: **plasmalemmasomes (plasmalemmasomes - tubes = plasmatubules)** the membranes of which are formed directly from the plasma membrane, and **lomasomes**, whose membranes are formed from the cytoplasmic membrane (Marchant, Robards, 1968). Vesicles that are produced by the Golgi apparatus or endoplasmic reticulum are related to lomasomes. They are usually incorporated into the plasmalemma and their purpose is to carry the materials for forming the cell wall. In certain situations they may exist in the form of bubbles within the entire space between the plasmalemma and the cell wall. Lomasomes, according to the results of studies performed during the 60-70's of the last century, may participate in the formation of the cell wall of some species of fungi and algae (Wilsenach and Kessel, 1965, Marchant e.a, 1967). The formation and distribution of plasmatubules refers to a process that may be associated with a modification of the emerging cell wall of higher plants (Marchant and Robards, 1968). Plasmatubule-vesicles and tubules (most often found at the same time) are found in autolysis of the protoplast between the primary cell wall and collapsing plasma membrane, as well as in normally functioning tracheal elements (Verbelen, 1977) and in differentiating xylem elements (Cronshaw, Bouck, 1965). Some authors argue that plasmatubules may play a role in the formation of the cell wall, for example, in collenchyma cells (Cox, Juniper, 1973). Other researchers believe the appearance of these structures to be an artifact of the fixation process (O'Brien, 1972). Some works present an entirely different view. For example, when using different fixation methods for plant material (barley) it was shown that plasmatubules were tubular invaginations of the plasma membrane and were formed between the apoplast and symplast where a high level of solute movement occurred. The appearance of the plasmatubules was independent of the fixation method used for fixing samples in the study of the ultrastructure of plant cells (Chaffey, Harris, 1985, Harris, Chaffey, 1985). In literature many examples of plasmatubules have been detected in different tissues of various plant species (Evert e.a, 1977, Kandasamy e.a, 1988, etc.). Their structure is partially described in earlier papers (Marchant, Robards, 1968). It is stated that plasmatubules represent the formation of plasmalemma, yet the structure of these formations is not yet known. The process of their formation (plasmatubules) just budding from the plasma membrane is problematic to describe, for it is not clear whether they have a plasma membrane or some other type of layer. The data accumulated during our research in the study of the secondary cell wall formation in plant cells of different species allows us to contribute to the debate on some of the reasons for plasmatubule formation in the area between the plasma membrane and cell wall, the dynamics of their formation and their structure.

The objective of our research on the formation of plasmatubules was:

1. To analyze the ultrastructure of the formations found between the plasma membrane and the secondary cell wall (during the formation of the polysaccharide cell wall of flax phloem fibers, tension wood of poplar, hairs of cotton seeds (trichome), secondary phloem fibers of hemp);

2. To determine the origin (the cause) of these paramural bodies.

The above objects have been selected with a purpose. First, they are classic examples for studying the formation of secondary cell walls. Second, the material we have obtained over the course of time in the study of the ultrastructure of the process on these objects allows us to comprehensively analyze the observed structure formation.

1. Results

1.1. Primary phloem bast fibers of flax stem (*Linum usitatissimum* L.).

The structure of primary phloem bast fibers of flax, which has been described in earlier papers by the author and colleagues (Sal'nikov e.a., 1993) is clearly visible on the cross and longitudinal sections of the stem of flax in the late stage of rapid growth. In this paper emphasis will be placed only on the formations called plasmatubules that have already been noted in the above mentioned article and that have been defined as plasmatubules. When considering these formations, we will continue, to a certain point, to call them plasmatubules. **Figures 1-6** present the stem sections including the sites of the primary phloem bast fibers of flax in cross sections and a longitudinal section along the whole length of the fibers at different magnifications. In all presented micrographs, plasmolysis in the cells of the primary phloem bast fibers of flax stem was observed to varying degrees.

It should be noted that the plants were fixed at the stage when the secondary cell wall formation process was in its most active phase. Plasmatubules were observed to be forming in the space between the plasmalemma and the cell wall, practically around the entire perimeter of the cell. The quantity, degree of branching and variety of plasmatubule forms also varied greatly. At some sites one could see individual plasmatubule formations in the form of bubbles, bordered by a double membrane (very rarely), visually similar in structure to the membrane of the plasmalemma.

1.2. Tension wood of *aspen (Populus tremula* x *tremuloides Michx.)*.

Gelatin wood fibers (tension wood) are a special type of xylem fiber that are formed in dicots in response to shoot bending (Mellerowicz, Gorshkova, 2011, Gorshkova, Biogenesis of plant fibers, 2009). The ultrastructure of xylem fiber in the tension wood is presented in cross section micrographs **(Figs. 7-13a)**, which show that at a strong plasmolysis when the thickness of the secondary cell wall is significant, numerous plasmatubules are observed in the space between the cell wall and the plasma membrane. When the protoplast is squeezed hard, the plasmatubules are evenly distributed throughout the formed "free" periplasmic space. The number and distribution of plasmatubules depends on the rate of plasmolysis. The size of the plasmatubules varies greatly (20 - 300 nm in length, 20 - 40 nm in diameter). In the bulk, short plasmatubules were observed, as well as numerous bubbles from which, apparently, the plasmatubules were assembled.

Since the micrographs obtained from the tissue sections present a static picture of what occurs therein, it is possible to consider not only plasmatubule assembly, but also their destruction. In those tissues where the degree of plasmolysis was weak, the plasmalemma appeared to be retreating from the cell wall to a slight distance, and plasmatubules were observed to be "huddled" near the membrane in small groups or as a single monolayer between the cell wall and plasma membrane. As a rule, in places across from the plasmatubules gathering plasmalemma we observed an actively functioning Golgi apparatus and a rough endoplasmic reticulum proliferation.

1.3. Cotton (*Gossypium hirsutum* L.) seed hairs (trichome).

In cotton seed hairs, fixed by the freeze-substitution method, we have found structures similar to plasmatubules, as described in the previous two cell types **(Figs. 14-17)**. Fixing plant tissues by freeze-substitution aims to preserve the protein localization for further immunocytochemical reactions (Salnikov e. a, 2001, 2003). The structure of organelles and membrane structures on sections obtained after such fixation ranks lower in quality to those fixed by the classical chemical method for studying the ultrastructure of the object itself. However, in our micrographs we found structures that we have designated as plasmatubules. The plasmatubule location (between the plasma membrane and cell wall), the

10

process under study (intensive formation of the secondary cell wall), the dynamics of formation (from individual vesicles and membranes by the formation of short cords of plasma membrane), are all similar to those described above for the previous cell types.

1.4. Secondary phloem fibers of hemp (*Cannabis*).

When investigating the dynamics of formation of the primary and secondary fibers of hemp stem, plasmatubules were also detected in cells exposed to plasmolysis **(Figs. 18-20)**. It should be noted that the presence of plasmatubules in hemp fibers (forming the secondary cell wall) in cells exposed to plasmolysis is not the norm, meaning the fibers fixed to study the ultrastructure using the electron microscopy. Apparently, this depends on the quality of fixing and especially on the time of penetration of fixatives used for electronic microscopy in tissue. The time of penetration, in its turn, depends on the tissue density, the maturity of the secondary cell walls and the depth of the tissue in a particular organ of the plant.

1.5. The shape and dimensions of the plazmatubules.

Plasmatubules are characterized by various forms **(Fig. 4, 4a, 6 and others)**. They have often been observed (in flax stem primary phloem bast fibers) as individual bubbles of different sizes, dumbbell-shaped formations, tubes of different lengths and diameters, segmented tubules like braids, twisted, branching tubules and others. The smallest plasmatubule diameter is comparable to the diameter of classical microtubules, 24-30 nm. However, its (diameter) maximum value can reach an order of magnitude greater than its minimum size. The plasmatubule diameters are more or less constant (between 20-30 nm), but there are also numerous structures (plasmatubule) that differ by 2-3 times from this size. It is logical to assume that the diameter of the "grown" plasmatubules depends on

11

the diameter of the bubbles, which are separate from the plasma membrane. The length of plasmatubules also varies considerably.

2. Discussion

It is very difficult to identify, trace (using chemical fixation) and understand the formation process of the largest, longest, branched plasmatubules. Does this vesicle fusion occur in short and thin plasmatubules and then aggregate into larger and more complex structures (which seems plausible), or does this process take place in the opposite direction?

We observed plasmatubule congestion in some areas at the perimeter of the cell, which is apparently associated with varying degrees of plasmolysis. As plasmolysis intensity increased, so increased the amount of plasmatubules. At the same time, the amount and density of the formed plasmatubules may be linked to the stage of cell wall formation, and to the type of cell wall being formed - whether primary or secondary, as each type has its own characteristics. For example, the cell wall types differ in terms of the composition of the polysaccharides that compose each one (Gorshkova e.a, 2003, Gorshkova, 2007).

Clutch, that is, the contact of the cell wall with the plasma membrane during the formation of the primary cell wall is apparently stronger than during secondary cell wall formation. The elasticity of the primary cell wall is higher than that of the secondary cell wall and therefore its resistance to impact, such that plasmolysis is manifested more efficiently (i.e. plasmolysis impact is lower).

The formation of plasmatubules was observed throughout the plasma membrane in the primary phloem bast fibers of flax stem. The question arises as to whether the plasmatubule formation is somehow connected with the formation of the secondary cell wall. We do not support this view regarding the objects studied in our work. In our experiments on the localization of the secondary cell wall components considered in this paper, more than 30 different antibodies for cell wall polysaccharides were studied, including for enzymes involved in the synthesis of polysaccharides, as well as for the individual monomers comprising the structure of the polysaccharides. Experiments for the study of reactions to the components of microtubules and actin filaments were also carried out. None of the immuno-cytochemical dyeing was observed in response to the plasmatubules (Salnikov e.a, 2001, 2003, 2008 and others). There is evidence (author suggests, more or less…) showing part of the plasmatubules in the cell wall formation, e.g. cells collenchyma (Cox and Juniper, 1973). We observed the classic processes (exocytosis), which were directly linked to the plasma membrane, followed by participation of some of its layers in the formation of the secondary cell wall. The results imply that exocytosis vesicles with their electron-dense contents secrete it directly into the cell wall during the formation of the secondary cell walls of the primary phloem bast fibers of flax stem (Salnikov e.a, 2010).

Plasmatubule formation is clearly seen on the surface of the plasma membrane in the micrographs. However, it is difficult to trace the dynamics of this process from static images. One might imagine that plasmatubule formation may proceed as follows **(see Fig. 6)**. During plasmolysis, small bubbles bud from the plasma membrane one by one, move into the space between the plasma membrane and cell wall, and then merge into tubes of varying lengths. We might also witness the collapse of the plasmatubules (or Hechtian strands) formed during plasmolysis. Hechtian strands, when torn, are transformed into the ranks of the chain-bubbles, or twisted series-chains (Domozych ea, 2003). One could draw an analogy to a thread torn by being pulled on horizontally. After the break, its ends are twisted too. Plasmatubule length, the rate of plasmatubule formation, their branching, and the presence of individual bubbles, all depend on the degree of compression of the plasma membrane, i.e. the degree of plasmolysis. The micrographs show clearly-defined structures of plasmatubules similar to the structure of microtubules. However, such microtubule structures are created by swirling the constituent proteins into a braid, whereas in the case of plasmatubules, the structures are formed via the assembly of small bubbles into long aggregates, which may be branched. There are those who believe that plasmatubules are also twisted tubes

(Kandasamy e.a, 1988). Branching of plasmatubules may occur due to some limit in possible length of plasmatubules, where at some point they breach the tube assembly, which leads to the formation of some chaotic, tubular or pseudo tubular structure. The membrane that borders the vesicles of budding plasmalemma differs from the plasmalemma membrane at least in its thickness. Moreover, it is difficult to say that plasmatubules are limited by a membrane, if one bears in mind the classic, three-layer structure observed after fixation in glutaraldehyde and osmium tetroxide. By what membrane they are limited? The membrane structure of plasmatubules includes phospholipids, the layer of which eventually borders the formed plasmatubule. In dense areas of plasmatubule formation where a flood of budding vesicles has occurred, a distinct change in the structure of the plasma membrane was observed **(Fig. 4, arrows)**. In this case, the plasma membrane was of diffuse structure, whereas in the surrounding areas we observed plasmalemma of the classic form, such as after standard chemical fixation.

2.1. Question: Are they plasmatubules?

The fact that plasmatubules, or structures similar to them in appearance, are formed from the plasma membrane is accepted without a doubt. However, the structure of bubbles (e.g., membrane), the assembly of bubbles in the tube (how), what determines the degree of merger, and the reasons for plasmatubule formation, remain unclear. We consider the formation of plasmatubules in the primary phloem bast fibers of flax during the active formation of the secondary cell walls. An intensive release of material occurs on the surface of the fibers via insertion of bubble-fringed membranes into the plasmalemma (Salnikov ea, 2010). A large amount of "excess" membrane is formed, which the cell under normal operating conditions gradually copes with. Traces of the intensely held exocytosis remain as wavy plasmalemma. During plasmolysis, which we assume occurs with fixing at our facilities, a compression of the protoplast at varying degrees is observed. Thus, a superposition of two processes occurs. On one hand, the plasmalemma receives a "surplus" of membrane vesicles, resulting in exocytosis, and on the other hand, there are "surplus" membranes during plasmolysis, abruptly and significantly increasing the number of "unnecessary" membranes. As a result, an avalanche formation of plasmatubules occurs. We have called this phenomenon a plasmatubular "explosion" (an "explosion" that is nothing but the result of plasmolysis). A similar picture of a plasmatubular "explosion" in similar processes occurring in the cell can be found in a number of previously published articles (Cronshaw, Benjamin, 1965, Moore, 1982).

We often observe the structures between the plasmalemma and cell wall in the form of bubbles or short tubules in the form of beads. We can explain this as a violation of the first stage of plasmolysis by the ongoing fixation. Plasmolysis occurs in the form of tension that arises between the plasma membrane and cell wall at certain points and areas where "anchoring" exists. At fixation this tension increases, but the formation of Hechtian strands is terminated because of a sudden stop of the process (when fixation is involved - this is the main task of fixation for the study of ultrastructures). This leads to the "shrinkage" of emerging Hechtian strands. If they are relatively short, then bubbles are found. If they are long enough, one may observe a chain of bubbles.

We agree that plasmatubules are not an artifact of fixation (Chaffey, Harris, 1985), to a certain extent. By this, we suggest that when the standard process of secondary cell wall formation exists, the "quiet" and well-known work of organelles is involved, then a number of structures like plasmatubules can occur (not always). With active Golgi vesicle exocytosis, a very wavy plasmalemma with numerous bumps and hollows is observed - the result of many embedded bubbles. Naturally, the plasma membrane strives to revert back its normal, smooth state. If a growing cell with a primary cell wall increases its volume quickly, the wavy cytoplasm is also able to quickly equalize. If something inhibits cell expansion or if it does not have time to compensate for the excess embedded membranes, we can observe two processes: First - compensatory endocytosis, where the

15

plasmalemma forms invagination or bud vesicles within the cytoplasm, forming multi-vesicular bodies and similar formations that are then lysed, or else remain until the end of the cell's life, and second - the formation of plasmatubules, the "outgrowth" of the plasma membrane by separating vesicles from the outer membrane of the cell or from the outer phospholipid layer. The issue regarding the "inferiority" of the membrane and the membrane of the plasmatubule bubbles being formed (if that process does indeed exist) is raised for the following reason. Very rarely does one see in the micrographs the classic three-layer membrane in plasmatubules (Hechtian strands) in any of the slices (after chemical fixation and or fixation of another kind). However, in bubbles forming plasmatubules, this membrane is observed much more frequently.

According to A.E. Vasiliev (personal discussion) "... these outgrowths out (meaning plasmatubules) – are the intussusception of plasmalemma. In cells, an intense centripetal deposition of the secondary containment occurs. The surface of area of the plasma membrane is reduced. It is known that excess of the plasma membrane resulting from the merger with the Golgi membrane vesicles during exocytosis and from reducing protoplast surface is adapted by the so-called compensatory endocytosis (separation of bubbles from the plasma membrane and their delivery into the cell). Apparently, the removal in this case is slower, the surface of the protoplast shrinks. As a result plasmalemma is forced to form folds". The plasmatubules observed in cotton-seed hairs were discussed in a personal conversation.

Some membranes undergo a high rate of vesicle production by the Golgi apparatus that leads to their embedding in the plasmalemma, and this process can significantly exceed the rate of cell growth. This leads to the formation of a huge number of the "real" plasmatubules (here we introduce the term "real" plasmatubules to be applied hereinafter). For example, such a phenomenon is observed with the growth of pollen tubes in vivo (Kandasamy e.a, 1988). This discrepancy in rates leads to plasmatubule formation in different configurations, their extensive crowding up to the formation of peculiar clusters and isolated storage on the perimeter of the cell. Note also that in the study of the ultrastructure of the objects described above one may find (in an amount not subject to statistical processing) plasmolyzed cells (plant cells subjected to plasmolysis), but shall not find plasmatubules in them. According to our observations, the presence of plamolyzed cells depends on many factors, including cell properties, state at any point in time, and the tissue types that are affected by the fixation process for the object under study. Apparently, in prepared fibers, there exist in varying degrees, long or very long cells for fixing (vessels). In further cross sections, we often viewed cells that had undergone mechanical damage (clipping of individual sections of the stem, and such). The frequency in which cells of "small" size were seen intact was much higher than that of "large" size. Also, the more flexible the primary cell wall, the stronger its grip on the plasma membrane. The absence, more exactly properly quickly terminating processes associated with the formation of the cell wall (an active short-Golgi vesicle exocytosis), all contribute to an effective response to "incorrect" fixation, and as a consequence to a sharp decrease in the probability of observing plasmatubules.

Plasmatubule elements, themselves plasmatubules, or significant accumulations of plasmatubules can be found quite often in almost any plant tissue slice when observed carefully with a microscope. When a cell is exposed to certain influences such as physiological solutions and clamps, these can cause varying degrees of plasmolysis: from micro plasmolysis to "massive heart attack." If real plasmatubules were associated with the formation of the secondary cell wall, the regularity of their presence in the cells would be higher by orders of magnitude. Moreover, data on the use of antibodies is extremely scarce, based on which it could be argued as to whether there is any material in plasmatubules going to build cell walls. A unique work exists (Cox, Juniper, 1973), whereby autoradiography demonstrated the presence of radioactivity in lomasomes during the formation of collenchyma cells. According to the authors, lomasomes participate in the distribution of pectin and hemicellulose in the cell wall. This assertion has since been questioned in other works of the same period (Juniper e. a., 1981) and has not been confirmed in subsequent publications.

The analysis of numerous observations and the data obtained lead to the conclusion that plasmatubules found in cells forming a secondary cell wall may be:

1. Mainly the result of plasmolysis caused by tissue fixation procedures that are conducted for the study of fixed tissue cells, via static pictures, of their ultrastructure and Hechtian strands.

2. Caused by an "ideal" fixation process. A small portion of plasmatubles may be formed due to the fact that the removal of membrane excess after active exocytosis "is slower than the shrinking of the protoplast surface. As a result plasmalemma is forced to form folds." (Vasiliev A.E. personal communication).

According to the literature, plasmatubules (real plasmatubules) are credited with the following functions and features:

1) A short-term modification of the plasma membrane – an increase of the surface area. Its properties are similar to those of transfer cells - specialized parenchyma cells of the phloem (Pate, Gunning, 1972, Harris ea, 1982) that are known to effectively absorb sugars and amino acids. A distinctive feature of transfer cells are the numerous outgrowths of the cell walls. Due to the outgrowths directed into the cells, the surface area of plasma membrane increases. At the same time the membrane capacity increases and a favorable environment is created for the absorption of substances;

2) Plasmatubules are involved in the processes are involved in secondary cell wall formation processess (Marchant, Robards, 1968);

3) Different types of plasmatubules exist (Verbelen, 1977);

4) Plasmatubules - intussusceptions of plasma membrane - appear where the flow of solutes between apoplast and symplast increases (Chaffey, Harris, 1985);

5) Formation of plasmatubules in pollen tubes is associated with high consumption and intensive transport of nutrients (sugars and amino acids) during the growth of the pollen tubes (Kandasamy e.a, 1988). **Pollen tube plasmatubules formed in the plasma membrane create a kind of network, a regular structure,**

having a recognizable system formation. Such plasmatubules are an example of structures that can be justifiably called real plasmatubules.

If "real" plasmatubules refer to actual existing operating entities for vital activity of whole cells (as organelles), including the ultrastructure and dynamics of the formation (e.g., in the case of barley squamous cell embryo (Harris e.a, 1982, Chaffey, Harris, 1985), maize leaf cells (Evert e.a, 1977), tobacco pollen tubes (Kandasamy e.a, 1988) and others , but do not refer to structures like Hechtian strands (being the result of convulsive plasmolysis), then according to some authors, they most likely perform a role similar to that characteristic of outgrowths of the plasma membrane in the cell wall in the transfer cells of plant phloem (Offer ea, 2002).

2.2. Plasmolysis and Hechtian strands.

Based on our observations, the conclusion follows that the observed structures are the result of plasmolysis, at least in the cells that form the secondary cell wall. Furthermore, they are found in most other cells in different tissues and under different conditions (according to the literature and our own observations). If we recall basic plant cell physiology and refer to the definition of plasmolysis, we find that plasmolysis is the separation of the protoplast from the cell wall.

Under hypertonic conditions, the cell loses water to the external medium and shrinks in volume. The plasma membrane eventually gets torn completely from the cell wall and continues to contract. This second stage of plasmolysis is called evident plasmolysis. As exosmosis continues, the shrinkage of the cell and cytoplasm reaches a minimum limit and no further shrinkage in volume is possible. Cytoplasm is completely free from the cell wall and remains in the center of the cell. Based on the final shape of the cytoplasm the final plasmolysis is divided into two types: concave plasmolysis and convex plasmolysis. In the former, the cytoplasm assumes a concave shape and in the latter it assumes a convex shape (Oparka e.a 1990, Oparka, 1994).

There exists a *corner-type* plasmolysis, in which the separation of the protoplast from the cell wall occurs in only some areas, a convex-type plasmolysis when large areas of plasma membrane are delaminated, and a *concave-type* plasmolysis, where complete plasmolysis occurs and communication between neighboring cells is almost completely destroyed. The c*onvex-type* plasmolysis is often reversible - in a hypotonic solution, the cells regain lost water, and deplasmolysis occurs. The *concave-type* plasmolysis is usually irreversible and leads to cell death. If the protoplast connection with the cell wall is conserved in some places, further decreases in the volume during the plasmolysis of protoplast becomes irregular. The protoplast remains connected to the shell by numerous Hechtian strands. This is called the *convulsive-type* plasmolysis (Hayashi e.a., 2003, 2006). **(Fig. 24)**.

As a result of plasmolysis between protoplast and the cell wall, filament strands are formed from the plasma membrane and remain in contact with the cell wall. These strands in foreign publications are called Hechtian strands. Hechtian strands were first described in the nineteenth century (Pringsheim, 1854, Nageli, 1855), but were eventually named after Hecht (1912), who also observed and described these formations. Bands seen between the plasma membrane and cell wall are found in various organisms, including algae, fungi and higher plants (references). They are also found in the cells of the callus (Buer, 1998, Buer ea, 2000). Despite the fact that the Hechtian strands occur only during plasmolysis and their dynamics are seemingly obvious (compensation of the plasma membrane surface during protoplast compression), little is known about the functions, structure (almost exclusively viewed via light microscopy) and physiological properties of these "strands". Hechtian strand diameters range from 30 to 250 nm (which corresponds to the size of the plasmatubules described by us and others), while their length is strongly dependent on the degree of plasmolysis. According to our observations, at very low intensity and short plasmolysis times when Hechtian strands just begin to form, if at this point the tissue is fixed for the examination of its ultrastructure, the plasmolysis process comes to an abrupt stop, which leads to the rupture of the Hechtian strands and to the formation of bubbles, short tubes or short strings of bubbles **(For example Fig. 4-6)**.

Hechtian strands are credited with the following functions and features according to the literature (Buer e.a, 2000):

1) Preservation of the excess plasma membrane resulting from plasmolysis (Oparka e.a, 1994);

2) Support of the protoplast polarity by limited adhesion of the plasma membrane with the cell wall at the poles of the cell (Pont-Lezica e.a, 1993);

3) Organization of deplasmolysis providing for the recovery of cell functions (Pont-Lezica e.a, 1993);

4) The horizontal internal Hechtian strands have esterase activity, which means that they contain the cytoplasm (Chang e.a, 1996);

5) Hechtian strands may contain elements of the endoplasmic reticulum (Operka e.a, 1994);

6) RGD - containing peptides (RGD - arginine-glycine-aspartate) inhibit the formation of the Hechtian strands and change the form of the plasma membrane in callus cells exposed to plasmolysis (Canut e.a, 1998);

7) Motor proteins associated with the granular endoplasmic reticulum (actin filaments, myosin) may provide the contact area of the plasma membrane with the cell wall and the subsequent formation of the Hechtian strands (Gunning, Steer, 1996);

8) Cellulose synthase complexes producing cellulose microfibrils of the cell wall may be localized on the membranes of Hechtian strands, which cause contact (anchoring) of the latter with the cell wall during plasmolysis (Lang e.a, 2004).

On the basis of these functions (which substantiate the value of Hechtian strand formation), one can assert not only that plasmatuble formation is a response of cells to plasmolysis, but also that these structures begin to perform certain functions including that of maintaining the integrity of the plasma membrane. The cell, before being placed in conditions causing plasmolysis, is in the process of growth or division, or the formation of the secondary cell wall. At plasmolysis the emerging Hechtian strands not only try to preserve the integrity of the cell, but also try to compensate by participating in cell function processes, those of its organelles and so on, to continue the ongoing process of development of the latter (cell).

2.3. Plasmatubules in growing pollen tubes.

The formation of a net of plasmatubules can be observed between the cell wall and the plasma membrane during the growth of pollen tubes (Kandasamy ea, 1988, **Fig. 26**). The activity of the plasmatubules in growing pollen tubes such as those of tobacco stems is associated with high nutrient consumption and the need for intensive transport of these nutrients during the rapid elongation of the growing pollen tubes. Moreover, the structure and organization of the plasmatubules, being plasma membrane invaginations, appears as a dynamic, organized, structured, and not chaotic system that is directly linked to the site of the actively occurring processes, in the absence of any external influence (pathology). Furthermore, the structure of such plasmatubules actually resembles a hollow tube even if its outer membrane appears undulating in sections. In contrast, in the "burst" Hechtian strands (the chain of bubbles) no internal channel is visible. As such, any statement as to the role of Hechtian strands as some type of channel with a transport function is, in our opinion, is very unlikely.

2.4. Lomasomes (Charosomes)

The formation of paramural bodies is observed in *Chara* (**Fig. 27**). Deep within the cells, lomasomes form bubbles with the participation of the Golgi apparatus, and possibly the endoplasmic reticulum. As noted in the book of Sedova T.V. (1977), aggregates of numerous, small, single-membrane-bounded vesicles are often observed in the folds of *Charales* plasmalemma when viewed both from the side facing the cell wall and the cytoplasm. Lomasomes appear in especially in large numbers during cell division, lying near the future cell walls. Based on the characteristics of the content and location it has been suggested that these structures are involved in the synthesis of the cell wall, in particular, proteins. The structures described as lomasomes were originally found in fungi, but later it was possible to identify their presence in algae, *Charales* for example, where they were originally named charosomes (Barton, 1965).

2.5. Invagination of the plasma membrane in the transfer cells.

In 1968, Gunning and his colleagues described a new type of modified cell companions – the transfer cells. These cells are located adjacent to sieve tubes. As a result of the additional irregular thickening of the cell walls, they form numerous internal protrusions, which constitute an almost tenfold increase in the area of the wall lining the plasma membrane. It is believed that such changes within the cells are connected with the need to actively absorb solutes from neighboring cells. Transfer cells have been detected in all plants. However, it is believed that active transport occurs in their absence as well (Gunning e.a, 1968, 1970, Offer e.a, 2002). **Fig. 28**.

Growths in the cell wall typical of transfer cells are not unique to this type of cell, as the need to boost the cell metabolism during the course of its growth and development can occur in very localized areas that vary greatly in size and area of the required increase of metabolite movement. Plasmodesmata play an important role in the transport of metabolites, but these are not the only structures involved in their transport.

The question arises as to whether the Hechtian strands begin to perform a function similar to that of the outgrowths of plasma membrane into the cell wall in the transfer cells, being the result of plasmolysis? We studied our objects during the most active period of secondary cell wall formation, when the receipt of cell wall components, particularly the matrix of polysaccharides, should be very intense. What happens to the active intake process at a time when a cell undergoes plasmolysis? Isn't "the immutability" of functions observed during this active period – the delivery of substances into the cell wall – by the replacement of structure (perhaps for the period until the cell copes with the consequences of the plasmolysis)? That is, the Hechtian strands can begin to fulfill the same function as that performed by outgrowths of plasma membrane in the cell wall in the transfer cells. Never before in the literature has a parallel been drawn to the comparison of the structure and function between plasmatubules and the Hechtian strands. The confusion in the interpretation of the ultrastructure of the "real" plasmatubules, the Hechtian strands and outgrowths of plasma membrane in transfer cells stems from the following:

1) Hechtian strands are mistaken as plasmatubules;

2) The outgrowths of plasma membrane in the cell wall in the transfer-specialized phloem bast parenchyma cells are mistaken as plasmatubules. What are these structures that are observed after the fixation of cells in the process of forming of the secondary cell wall: plasmatubules, Hechtian strands or formations with the "limited" functions typical of transfer cells?

2.5.1. Hechtian strands in primary phloem of bast fibers in flax stem.

On the basis of our own data analysis, we believe that the observed tubular formations that in earlier articles were identified by us as plasmatubules between the plasma membrane and cell wall in the phloem bast fibers of flax are no more than Hechtian strands. The static picture we see in micrographs is a consequence of cell fixation, mostly the broken Hechtian strands (many bubbles, drums and tubes of varying length and diameter). The individual Hechtian strands were preserved and stretched from the plasma membrane to the cell wall.

2.5.2. Hechtian strands in the fibers of tension wood of poplar stem.

At a high degree of plasmolysis when cytoplasm is extremely compressed, the remains of the destroyed Hechtian strands are uniformly scattered in the periplasmatic space of the xylem cells in a kind of gelatinous fiber. Often, very localized congestion is observed, most likely of vesicles and short tubules skintight both to plasmalemma and the cell wall. The ultrastructure is not characterized by highly elongated tubes, similar to Hechtian strands or "real" plasmatubules. Part of the vesicles and tubules "bite" cell wall.

2.5.3. Hechtian strands in cotton-seed hairs.

During the period of active formation of the secondary cell walls in the hairs of cotton seeds, we observed the accumulation (localized, not along the entire length of the plasma membrane) of vesicles and structures resembling ducts between the plasma membrane and cell wall. At some sites we noted the detachment of the plasma membrane from the cell wall. Interestingly, a portion of the detected tubules and vesicles remained "mounted" in the cell wall (this could be the result of fixing by freeze - substitution), that is, a portion remained in contact with the plasma membrane.

If the fixation process has caused plasmolysis for any reason, such as due to the fixing solutions, or due to bad fabric impregnated with resin, and plasmalemma separate from the cell wall, and structures similar to Hechtian strands or plasmatubules remain in the cell wall, then they are most likely related to real plasmatubules performing a role similar to membrane ramifications in transfer cells. If plasmatubules remain in the periplasmic space and maintain contact mainly with the plasma membrane, then they are likely to be Hechtian strands.

2.5.4. Hechtian strands in the secondary phloem fibers of hemp (*Cannabis*).

The remains of Hechtian strands are observed scattered around the perimeter of the cells in the periplasmic space in the secondary fibers of hemp stalk. The amount of these residues (individual tubes of varying lengths, bubbles) varies depending on the degree of plasmolysis. The observed clusters include "destroyed" Hechtian strands in the corners of the cells. Individual Hechtian strands of great length are found very locally in cells.

2.5.5. Vacuolar tubules.

During our observations of the dynamics of plasmatubule formation we noted the formation of such structures within vacuoles located in the cytoplasm of an active primary phloem cell of flax stem bast fiber **(Fig. 2, 3,10, 14, 20)**. The formation process of vacuolar tubules can be described similarly to the formation of plasmatubules. Vacuolar ducts are found in the form of separate bubbles of various sizes, dumbbell-shaped formations, tubes of different configurations and diameters, segmented ducts, branching tubules and other forms. It should be noted that the length of a vacuolar tube is shorter than that of plasmatubules, while the vacuolar diameter is 2-3 times larger than that of plasmatubules. We associate the formation of vacuolar tubules with the effect of plasmolysis, with a sharp contraction of the cytoplasm that leads to compression of the vacuoles within. As a consequence - budding of vesicles occurs inside the vacuole – a compensatory endocytosis. Vacuolar tubules are not paramural bodies as they are formed directly from the tonoplast membrane and accumulate in the "free" volume of vacuoles.

While studying the ultrastructure of wood fiber xylem tension, we found vacuoles containing structures - vacuolar tubules – in the cell cytoplasm, similar to those found in vacuoles and primary phloem of flax stem bast fiber. In a series of photomicrographs one can trace the stages of formation of the individual bubbles, which are likely to participate in the formation of vacuolar tubules. Moreover, in the cytoplasm of cotton-seed hairs we found vacuoles that were undergoing a similar formation of vacuolar tubules to that described in the two objects above.

The dynamics of the formation of vacuolar tubules in physiological nature is the result of compensatory endocytosis where the excess plasma membrane penetrates the interior of the cell. This phenomenon is observed in the intensive exocytosis in secretory cells. The phenomenon of compensatory endocytosis is also noted during plasmolysis in plant cells (Šamaj e.a, 2004). A similar picture of the formation of tubular structures in cells was observed on the inner side of the membrane tonoplast where plants were grown in a high salt content medium (Golombek e.a, 1994). The authors suggest that the appearance of tubular structures in the central vacuole of plant cells grown in a saline culture medium is due to the fact that these structures somehow control the transport of Na + and Cl-.

Thus, the formation of vacuolar tubules is the result of compensatory endocytosis, i. e. the formation of tubular structures and bubbles in the central vacuole and small cell vacuoles is the cell's response to a sharp contraction of the cytoplasm.

Conclusion

We believe that the formations, which have been previously thought to be plasmatubules in the objects under consideration are Hechtian strands without vacuolar tubules. Can Hechtian strands be attributed to paramural bodies? Based on the definition of paramural bodies, all correspondence, that is – yes. All structures were found between the plasma membrane and cell wall and were attached to a membrane, although the usefulness of the membrane we observed in these formations gives rise to many questions. If all these formations in our specific cases are the result of plasmolysis, it remains to be determined how to assign them the most appropriate name: Hechtian strands, plasmatubules, plasmalemmasomes, plasma membrane invaginations or some other name, and whether these definitions are equivalent or not.

In the cases we examined during the formation of the secondary cell wall we have dealt with structures that we believe have arisen as a result of plasmolysis. What function do these structures perform and begin to perform? We believe it depends on the situation. In the objects of the study the secondary cell wall is undergoing intense formation, when the cell mobilizes all available opportunities to continue the process, even using destructive changes, such as changes of the plasma membrane. To some extent, we can observe the cell's struggle for the restoration of the normal course of the formation of the cell wall. With the significant changes taking place in the structure of the plasma membrane, the paramural bodies that are formed may participate to some extent in the formation of the cell wall or in trying to deter further deterioration or slowing of its normal flow.

On the basis of literature data and the analysis of our numerous results, namely, the data on the structure of the plasma membrane formation under consideration, we conclude that the name plasmatubules is often incorrect and does not reflect reality, as the cause of their formation is often plasmolysis.

Consequently, plasmatubules – for the most part in plant cells, are Hechtian strands in one form of expression or another. In this case, by expression we mean the degree of filling of space between the plasmalemma and the cell wall by the Hechtian strands. This expression is primarily due to the degree of plasmolysis, and secondly, it depends on the type of plant tissue and how the cells of that particular tissue are able to withstand the impact of plasmolysis. It should be noted that the function of the Hechtian strands is most likely not limited to retarding the cell compression as a result of plasmolysis. In certain cases, these structures can perform specialized functions, leading to increased metabolism between cells that have suddenly found themselves in an emergency situation. Furthermore, there are the "real" plasmatubules, for example, pollen tubes of different plants. Despite the different reasons for the formation of these structures in plant cells (the Hechtian strands, "real" plasmatubules) and the functions they perform, they can both be called paramural bodies.

Thus plasmatubules (plasmalemmasomes) and the Hechtian strands are paramural bodies.

Additional experiments are required. In particular, immunocytochemistry studies using antibodies for specific components of the plasma membrane, the mathematical treatment of details of the plasmolysis process and study of the structures which form the vast network of the Hechtian strands (how a cell copes with the formation of such mass, the "defective" membranes) considering the steps, types of plasmolysis, etc. In the mathematical treatment of the data it is necessary to take into account the characteristics and properties of the classical membrane and the possibility of its modification. In this work, we used mainly our own data, so the conclusions drawn here may be insufficient and as yet, somewhat superficial, considering the complexity and breadth of the issue in question.

References

Горшкова Т. А., М. В. Агеева, В. В. Сальников, Н. В. Павленчева, А. В. Снегирева, Т. Е. Чернова, С. Б. Чемикосова. Стадии формирования лубяных волокон *Linum usitatissimum* L. // Бот. журн.- 2003.-Т 88.-N 12.-С. 1-11.

Биогенез растительных волокон. М.: Наука, 2009. 264 с.

Горшкова Т.А. Растительная клеточная стенка как динамичная система. М.: Наука, 2007. 429 с.

Сальников В.В., Агеева М.В., Юмашев Н.В., Лозовая В.В. Ультраструктурный анализ лубяных волокон в онтогенезе растений льна. Физиология растений. 1993, Т.40, N3, С. 458-464.

Сальников В.В. Структурно-функциональная характеристика формирования клеточных стенок в растительных тканях, специализированных на интенсивном синтезе целлюлозы. Автореф. дисс. д.б.н., Казань, 2005, 48 с.

Томилин Н.В., Васильева Е.В, Гиляров А.В., Черняк Т.Ф. Нарушения везикулярного цикла в синапсах ретикулоспинальных аксонов миноги после микроинъекции ГТФуS // Цитология. - 2007. - Т. 49, N 8. - С. 671-679.

Ageeva M., Kieft H., Lhuissier F., Vos J., Gorshkova T., Emons A., van Lammeren A. (2001) Microtubule cytoskeleton in elongating bast fiber cells of flax (*Linum usitatissimum* L.). Abst. 16 th Europ Cytoskeleton Forum Meeting 22-26 August 2001, Maastricht.

Barton R. 1965. Electron microscope studies on surface activity in cells of *Chara vulgaris*. Planta, 66, 2, : 95-105.

Chaffey N. J. and Harris N. Plasmatubules: fact or artefact? Planta 1985 Volume 165, Number 2, 185-190.

Cole, K., and S-C. Lin. 1970. Plasmalemmasomes in sporelings of the brown alga *Petnlonia debilis*. Can. J. Bot. 48: 265-268.

Cox G.C. and Juniper B.E. 1973. Autoradiographic evidence for paramural-body function. Nature New Biology, 243, 116-117.

Cox G.C., Sanders F.E., Tinker P.B. and Wild J.A. 1975. Ultrastructural evidence relating to host-endophyte transfer in a vesicular-arbuscular mycorrhiza. In: F.E. Sanders, BH. Mosse and P.B. Tinker (eds) Endomycorrihzas. Academic Press, London, New York, San Francisco. pp 297-312.

Cronshaw J., Bouck J.B. The fine structure of differentiating xylem elements Published March 1, 1965 // JCB V. 24 N. 3 p.415-431.

Domozych, D.S., Roberts, R., Danyow, C., Flitter, R., Smith, B. and Providence K. (2003) Plasmolysis, Hechtian strands formation and localized membrane-wall adhesions in the desmid, *Closterium acerosum (Chlorophyta)*. Journal of Phycology 39:1-18.

Evert R. F., Eschrich W., Neuberger D. S. and Eischhorn S. E 1977. Tubular extensions of the plasmolemma in leaf cells of Zea mays L. Planta 135, 203-5.

Golombek S.D., Hecht-Buchlor Ch., Ludders P. Ultrastructure of the endodermis of fig root tips in response to salinity 1994 Angrew. Bot. 68, 79-82.

Gorshkova, T. A., S. B. Chemikosova, V. V. Salnikov, M. C. McCann, V.V. Lozovaya, and N. C. Carpita. 1996a. Flax: a genetic and biochemical model to study phloem fiber development // Proc. of the Plant Polysaccharides Symposium (P. Colonna et al., eds.), Nantes, France, p. 160.

Gorshkova T.A., Chemikosova S.B., Salnikov V.V., Chen L.-M., McCann M.C., Lozovaya V.V., Carpita N.C. Flax: a genetic and biochemical model to study primary and secondary wall biogenesis // Proc. of 4th Workshop of the FAO Network on Flax (eds. P.Sultana et al.), Rouen, France, 1996 b.-P.457-462.

Gorshkova T.A., Wyatt S.E., Salnikov V.V., Gibeaut D.M., Ibragimov M.R., Lozovaya V.V., Carpita N.C. Cell-wall polysaccharides of developing flax plants // Plant Physiol.-1996 c.-V.110.-P.721-729.

Gorshkova T. A., Salnikov V. V., Pogodina N. M., Chemikosova S. B., Yablokova E. V., Ulanov A. V., Ageeva M. V., J. E. G. van Dam, Lozovaya V. V. Composition and distribution of cell wall phenolic compounds in the flax (*Linum usitatissimum* L.) stem tissues. Ann. Bot.-2000.-V.85.-P.477-486.

Gorshkova T.A., Salnikov V.V., Chemikosova S.B., Ageeva M.V., Pavlencheva N.V., van Dam J.E.G. Snap point: a transition point in *Linum usitatissimun* bast fiber development, 2003, Ind. Crops and Prod., N18, P. 213-221.

Gunning B. E. S., Pate J. S., and Briarty L. G. Specialized "transfer cells" in minor viens of leaves and their posible significance in phloem traslocation J.Cell Biol, June 1, 1968, V. 37: C. 7-12.

Gunning B. E. S., Pate J. S., Green L. W. Transfer cells in the vascular system of stems: Taxonomy, association with nodes, and structure. Protoplasma, Volume 71, Numbers 1-2 (1970), 147-171.

Gunning, B.E.S. and Steer, M.W. (1996). *Plant Cell Biology*, Jones and Bartlett: Melbourne. (ISBN 0867205040) 120 P.

Haigler C.H., Cai W.X., Gannaway J.G., Grimson M.J., Hequet E.F., Holiday S.A., Huang J., Jaradan T.T., Jividen G.J., Krieg D.R., Martin K.L., Nagarur S., Salnikov V.V., Strauss R.E., Tummala J., Wan C.H., Wu C., Wyatt B.G., Zhang H. Optimazing Secondary Wall Synthesis in Cotton Fibers. In: Proceeding of Cotton Inc. Conference in Genetic Improvement of Cotton, Dec. 5-6, 2000, San-Antonio, TX, pp. 147-165.

Harris**, N.**, 1981**:** Plasmalemmasomes in cotyledon leaves of germinating *Vigna radiata L.* (mung beans). Plant, Cell and Environment 4(2): 169-175.

Harris N., Chaffey N. J. 1985 Plasmatubules – real modifications of the plasmalemma. Nordic Journal of Botany Volume 6, Issue 5, pages 599–607.

Hayashi T., Harada A., Sakai T., Takagi S. $Ca2+$ transient induced by extracellular changes in osmotic pressure in Arabidopsis leaves: differential involvement of cell wall-plasma membrane adhesion. Plant Cell Environ. 2006 Apr; 29 (4):661-72.

Hayashi T, Takagi S. $Ca2+$-dependent cessation of cytoplasmic streaming induced by hypertonic treatment in *Vallisneria* mesophyll cells: possible role of cell wall-plasma membrane adhesion. Plant Cell Physiol. 2003 Oct; 44 (10):1027-36.

Hecht K. (1912) Studien über den Vorgang der Plasmolyse. Beitr Biol Pflanz 11:137–191.

Juniper, B.E., J.R. Lowton, and P.J. Harris. 1981. Cellular organelles and cell-wall formation in fibres from the flowering stem of *Lolium temulentum* L. New Phytol. 89: 609-619.

Kandasamy M. K., Kappler R. and Kristen U.. Plasmatubules in the pollen tubes of *Nicotiana sylvestris*. Planta 1988 Volume 173, Number 1, 35-41.

Lang-Pauluzzi,I.&Gunning,B.E.S.2000.Aplasmolytic cycle: the fate of cytoskeletal elements.Protoplasma212:174–85.

Lang-Pauluzzi, I. 2000.The behaviour of the plasma membrane during plasmolysis: a study by UV microscopy. J. Microsc. 198:188–98.

Lang I, Barton DA, Overall RL. Membrane-wall attachments in plasmolysed plant cells. Protoplasma. 2004 Dec; 224 (3-4):231-43. Epub 2004 Dec 22.

Lozovaya V.V., Gorshkova T.A., Grodzovskaya T.S., Ibragimov M.R., Zabotina O.A., Salnikov V.V. Cell wall formation in long-fibred flax stem tissues // Abstr. 6th Cell Wall Meeting (eds. M.M.A.Sassen et al.), Nijmegen, the Netherlands, 1992.-P.136.

Marchant, R., Robards, A.W. (1968) Membrane systems associated with the plasmalemma of plant cells. Ann. Bot. 32, 457-471.

Mellerowicz E.J, Gorshkova T.**A.** Tensional stress generation in gelatinous fibres: a review and possible mechanism based on cell-wall structure acomposition. J.Exp.Bot:2012,63:551-565.

Munns R. 2002. Comparative physiology of salt and water stress. Plant, Cell & Environment, V. 25., P.239-250.

Nägeli C. Pflanzenphysiologische Untersuchungen. Heft 1, Schulthess, Zürich1855135 (Quoted in Bower, 1883).

O'Brien T. P., 1972. The cytology of cell-wall formation in some eukaryotic cells. Bot. Rev. 38, 87-118.

Oparka K.J, Prior D.A, Harris N. Osmotic induction of fluid-phase endocytosis in onion epidermal cells. Planta. 1990 Mar;180(4):555-61. doi: 10.1007/BF02411454.

Oparka K.J. Plasmolysis: new insights into an old process. New Phytol. (1994), 126, 571-591.

Pringsheim N. Untersuchungen über den Bau und Bildung der Pflanzenzelle. Abt 1: Grundleinen einer Theorie der Pflanzanzelle, Hershwald, Berlin. 1854. pp. 12–13. (Quoted in Bower, 1883).

Salnikov V.V., van Dam J.E.G., van Hazendonk J.M., Lozovaya V.V. Microscopic studies of the cell wall formation of flax bast fibers. In: 4th European Regional Workshop of Flax, Roen, France, September 25-28, 1996, p. 463-466

Salnikov, V.V., Van Dam, J.E.G., Lozovaya, V.V.1998. Microscopy of cell wall formation in flax bast fiber. In. Natural Fiber, 2, 187-194.

Salnikov V., Grimson M., Delmer D., Haigler C. The localization of cellulose-synthesis associated sucrose synthase is changeable in cotton fibers. 9th International Cell Wall Meeting 2001, Toulouse, France, 2-7 September, p. 53.

Salnikov V., Grimson M., Seagull R., Haigler C. The localisation of sucrose synthase and callose in freeze substitution secondary- wall stage cotton fibers. Protoplasma, 2003, Jun; 221(3-4): P. 175-184.

Salnikov VV, Ageeva MV, Gorshkova TA. Homofusion of Golgi secretory vesicles in flax phloem fibers during formation of the gelatinous secondary cell wall. Protoplasma. 2008; 233(3-4):269-73.

Sardar H.S., Yang J., Showalter A.M. Molecular Interactions of Arabinogalactan Proteins with Cortical Microtubules and F-Actinin Bright Yellow-2 Tobacco Cultured Cells. Plant Physiology, December 2006, Vol. 142, pp. 1469–1479.

Jozef Šamaj, František Baluška, Boris Voigt, Markus Schlicht, Dieter Volkmann, and Diedrik Menzel Endocytosis, Actin Cytoskeleton, and Signaling[1] Plant Physiol. v.135(3); Jul 2004 PMC519036.

Verbelen J. P., 1977. A tubular type of plasmalemmasome in xylem of Phaseolus. Protoplasma 93, 363-7.

Volgger, M., Lang, I., Ovecka, M., Lichtscheidl, I. Plasmolysis and cell wall deposition in wheat root hairs under osmotic stress. Protoplasma. 2010 Jul; 243 (1-4): 51-62

Wilsknach R., Kessel M., 1965. The role of lomasomes in wall formation in Pemcillium vermiculatum. J. Gen. Microbiol. 40, 401—4

ABBREVIATIONS:

BR - bridges; CW - cell wall; GA - Golgi apparatus; G - gelatinous layer of cell wall; GV - Golgi vesicles; L - lipid body; M - mitochondria; CHL – chloroplast; CML- compound middle lamella; CW - cell wall; MT - microtubules; N - nucleus; NS - nucleolus; NE - nuclear envelope; NP - nuclear pores; HS - Hechtian strands; PM - plasmalemma; PL- plastids; PS - polysomes; PT - plasmatubules; PR - protective layer; R - ribosomes; RER - endoplasmic reticulum; SER - smooth endoplasmic reticulum; SP - simple pits; S1, S2 - cell wall layers; TP - tonoplast; TW - tension wood; V – vacuole, VT - vacuolar tubules. Scale bar: 1.0 μm

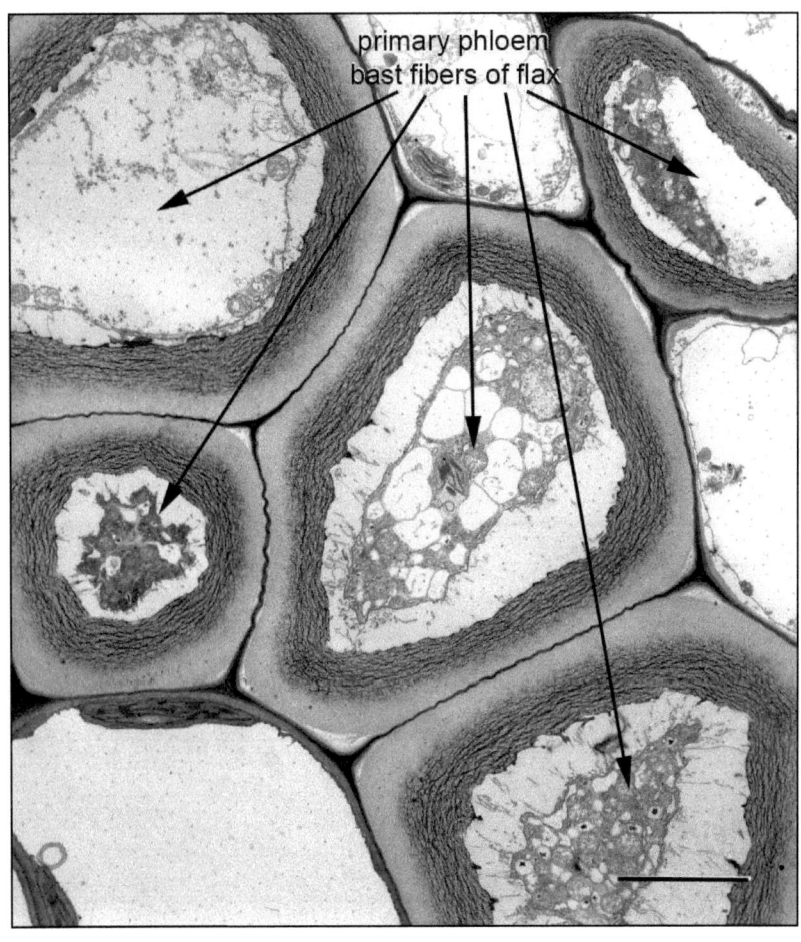

Fig. 1. Cross section of the bast fiber of flax stem (*Linum usitatissimum* <u>L.</u>).

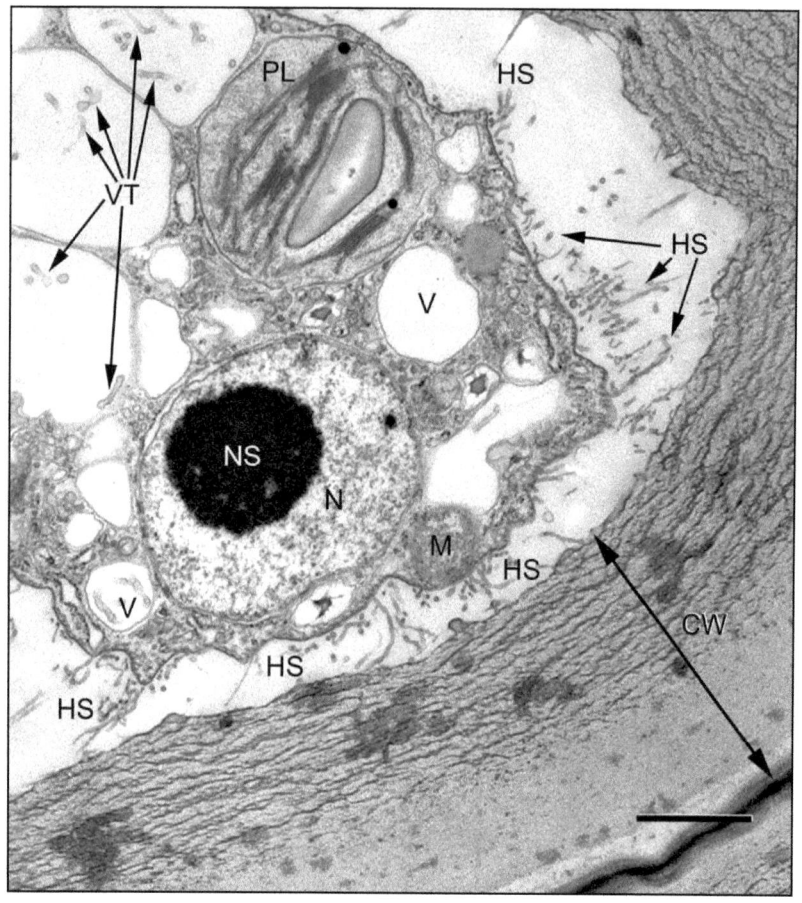

Fig. 2. Cross section of the bast fiber of flax stem. Ultrastructure of the bast fiber cell: plasmolysis, Hechtian strands, vacuolar tubules, condensation of the cytoplasm, accumulation of organelles in a limited space.

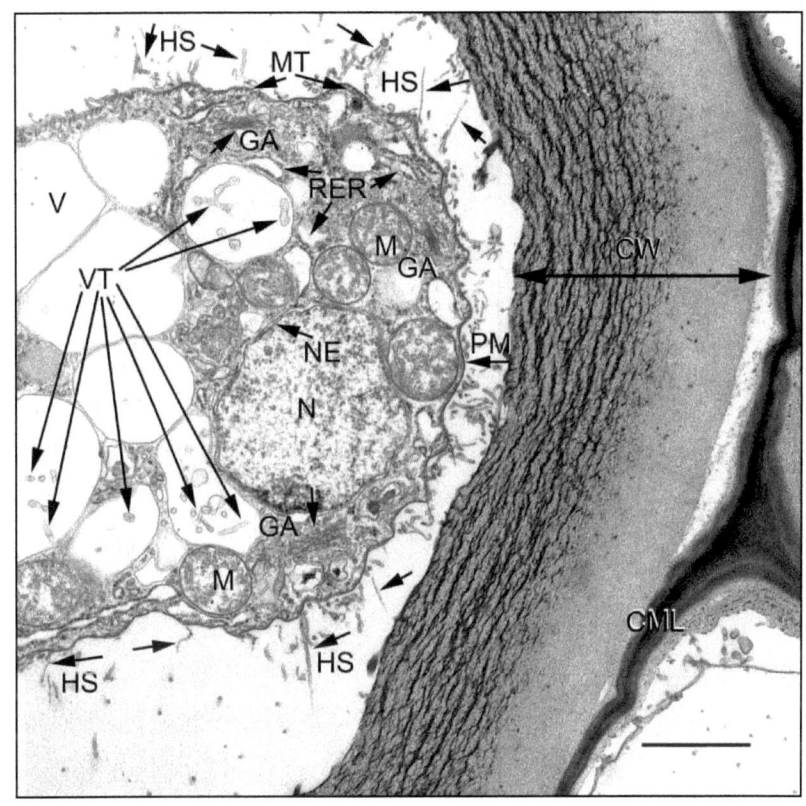

Fig. 3. The same.

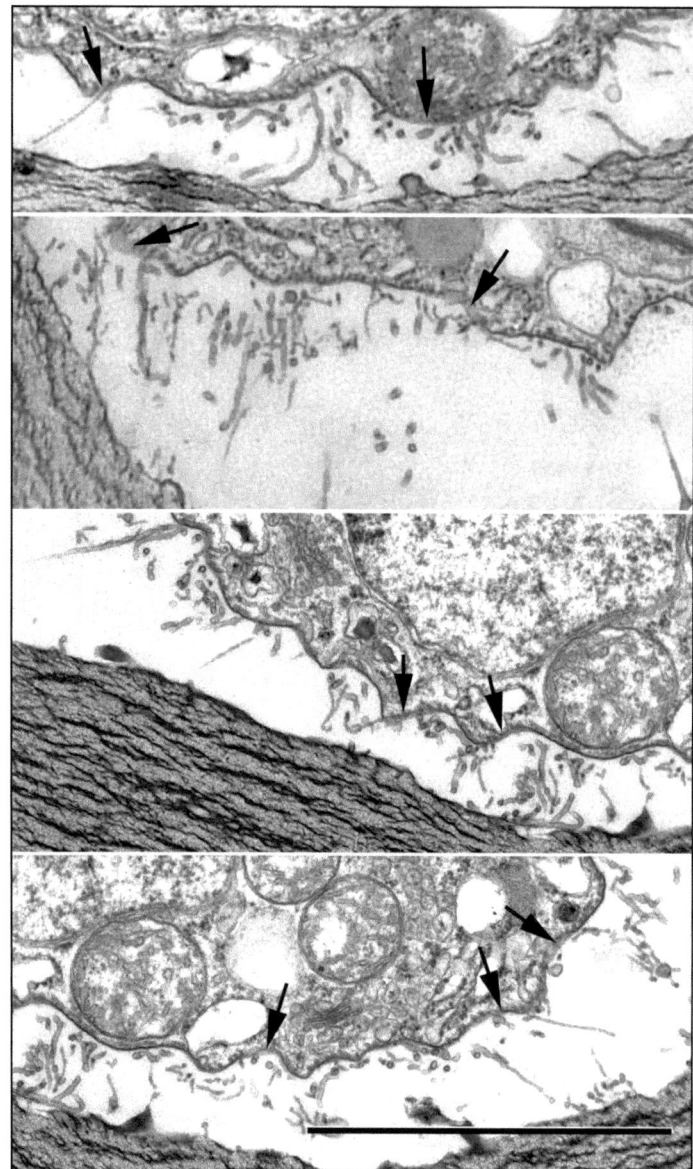

Fig. 4. The same in detail.

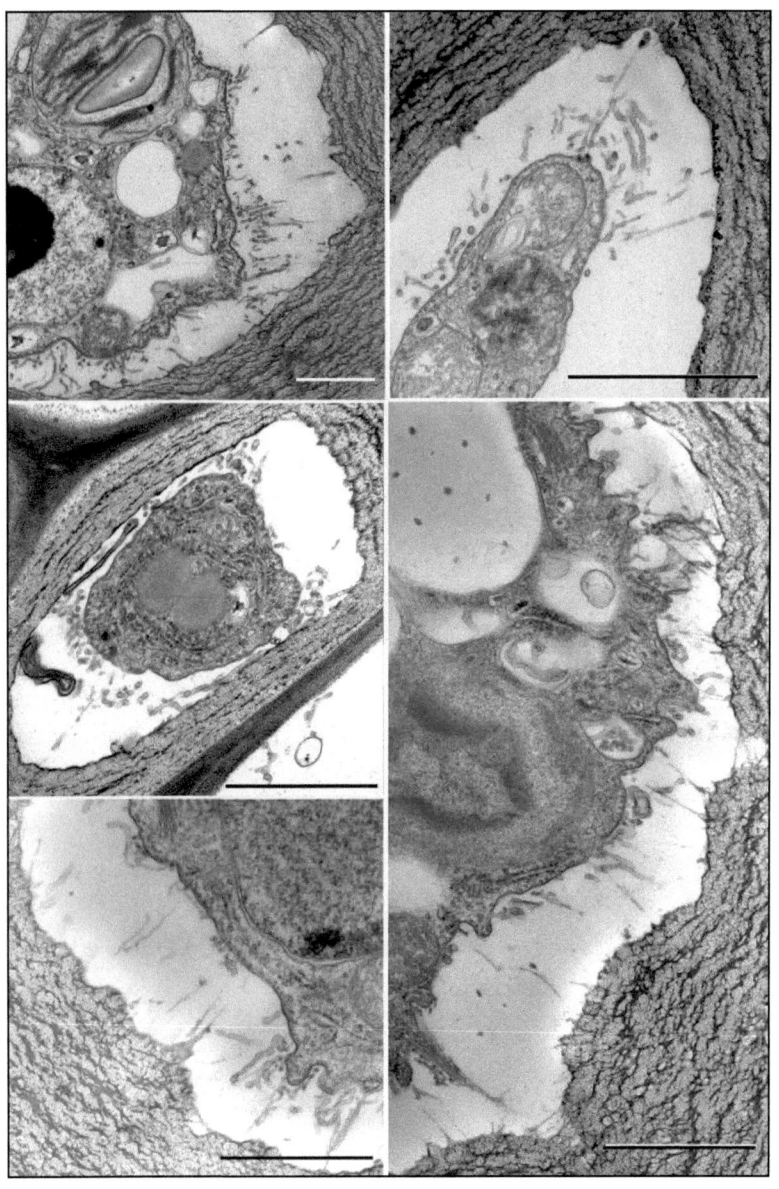

Fig. 4a. The same in detail.

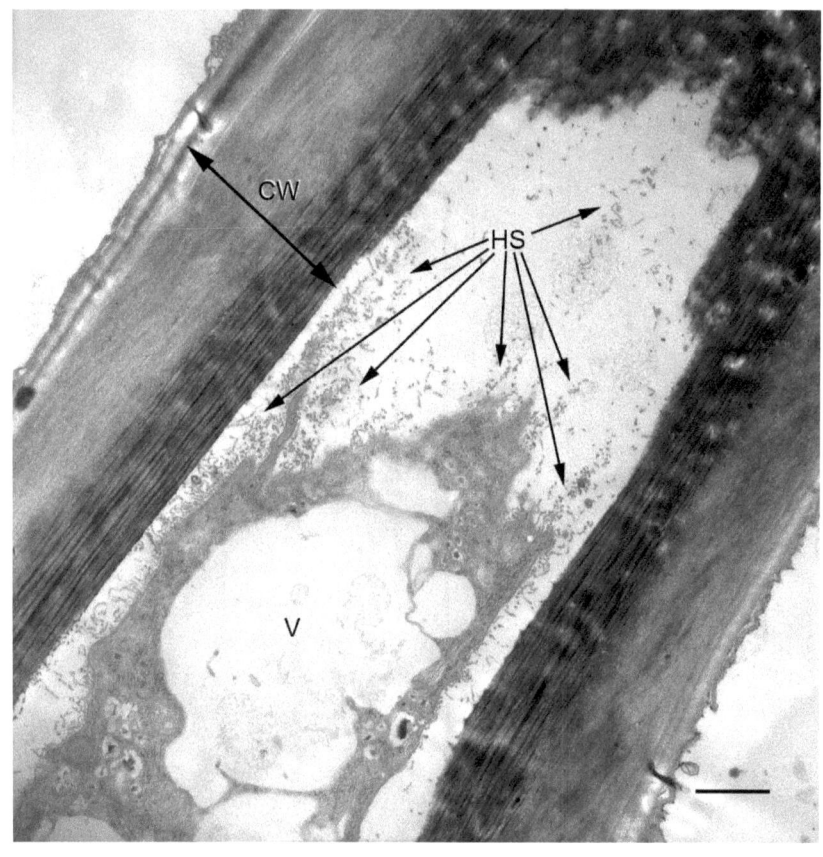

Fig. 5. Longitudinal section of the bast fiber of flax stem. Plasmolysis, multiple "burst" Hechtian strands.

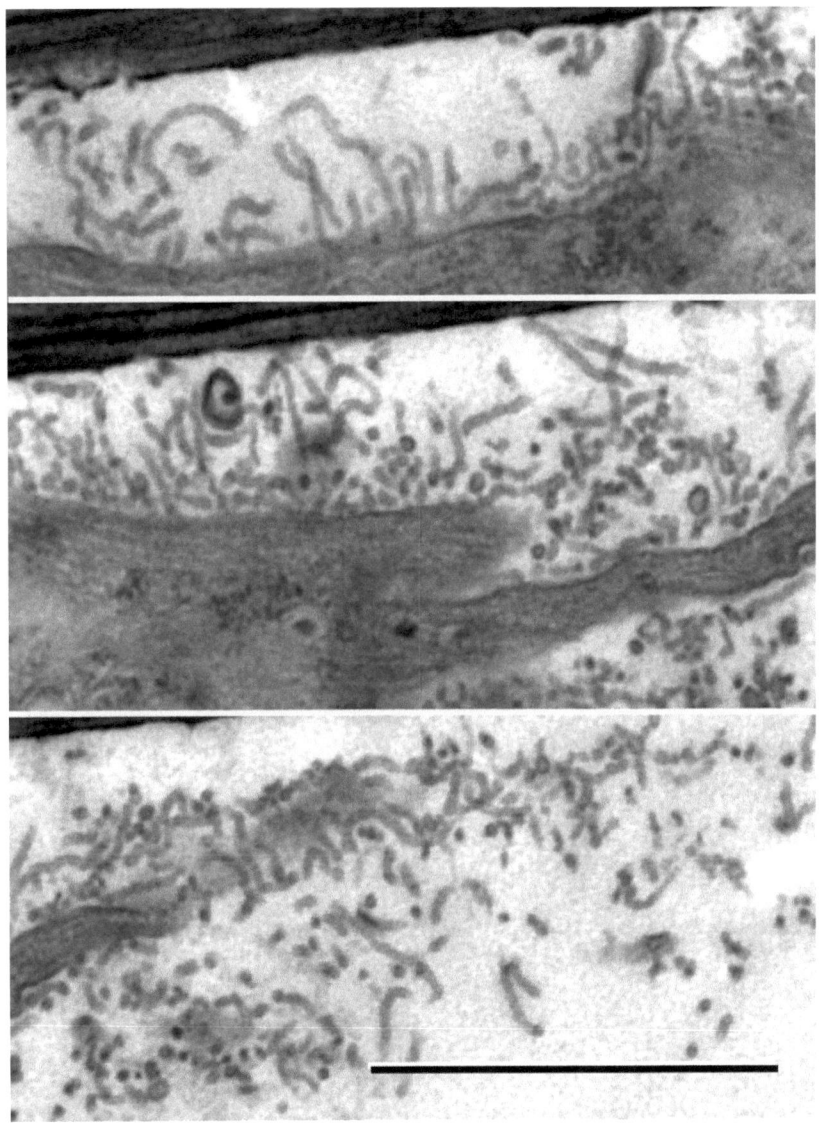

Fig. 6. The same in detail at high magnification.

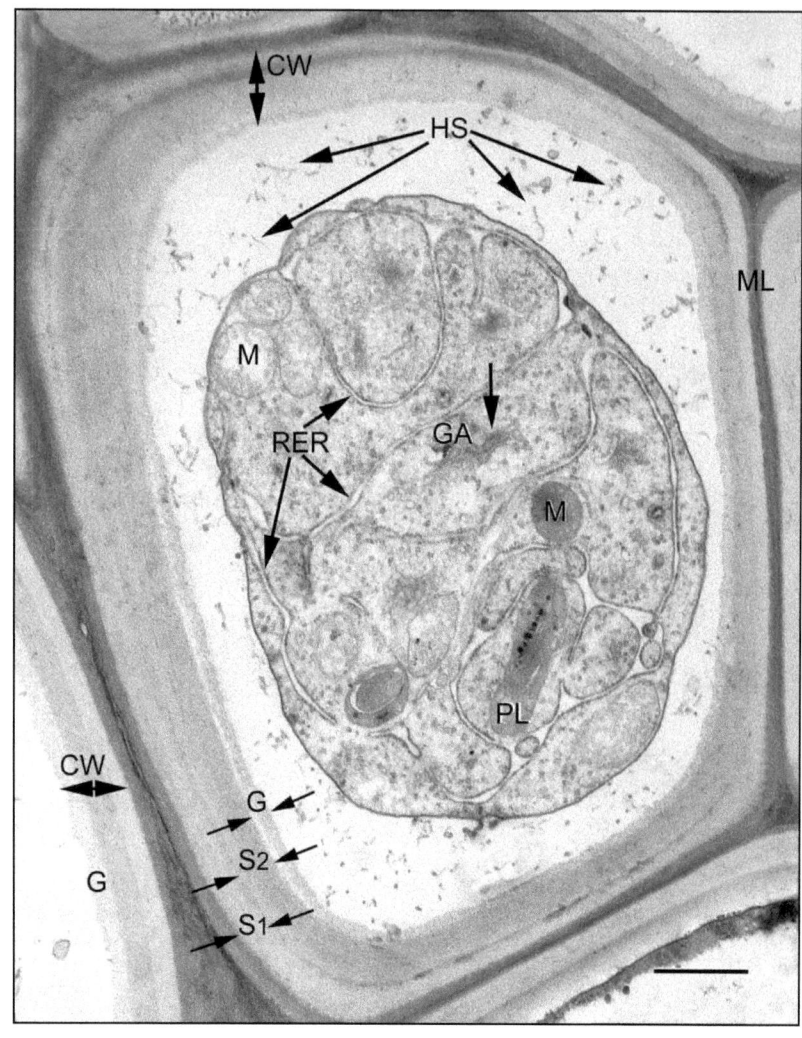

Fig. 7. Cross section of the xylem gelatinous fibers in tension wood in aspen stem *(Populus tremula L. x tremuloides Michx.)*. Plasmolysis, multiple "burst" Hechtian strands, condensation of the cytoplasm, accumulation of organelles in a limited space.

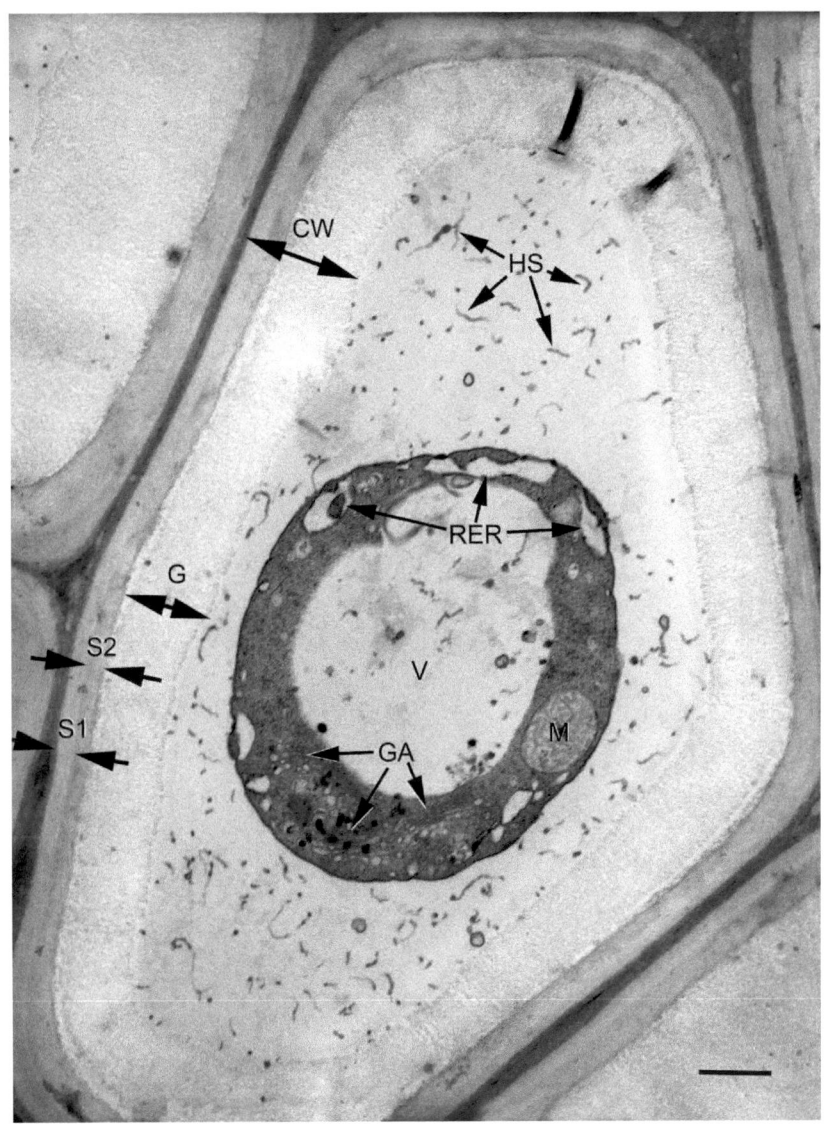

Fig. 8. The same. Numerous structures of the Golgi apparatus.

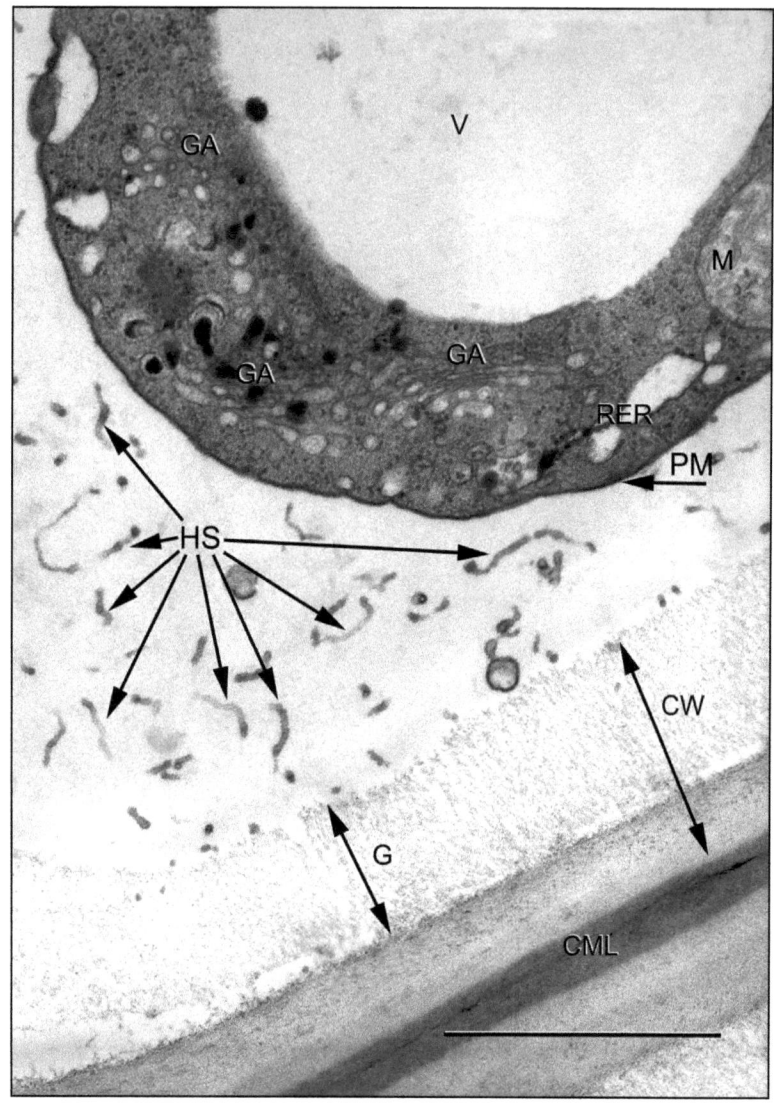

Fig. 9. The same in detail at high magnification.

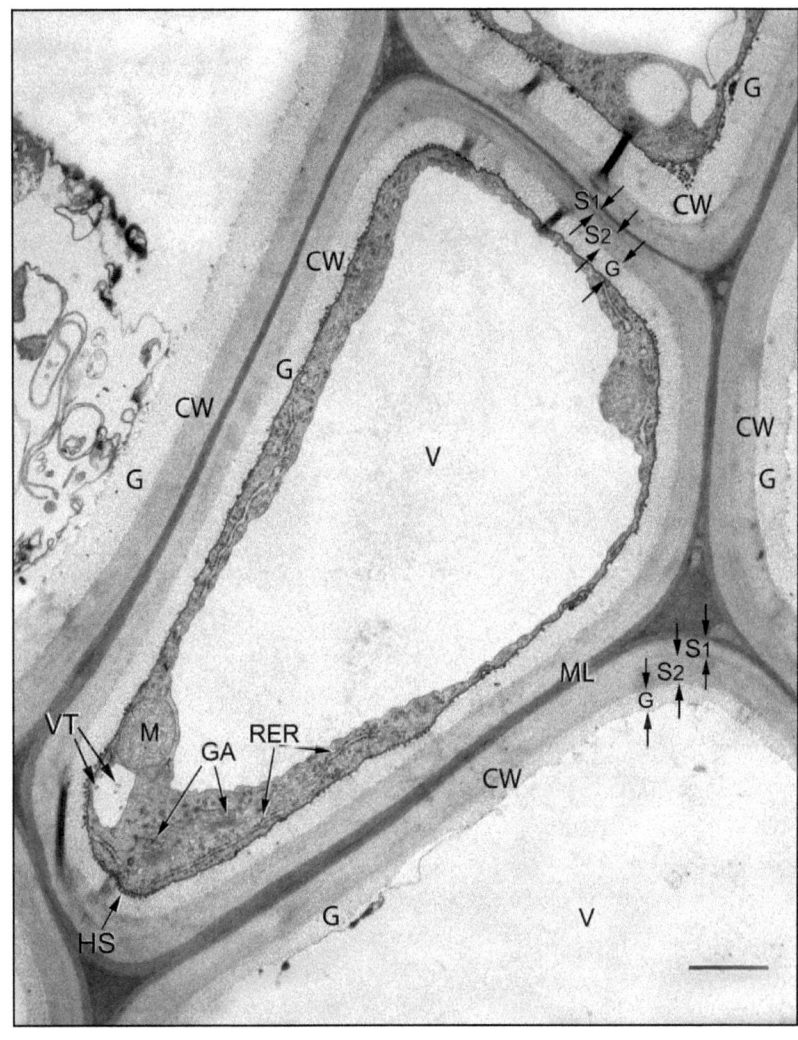

Fig. 10. Cross section of the xylem gelatinous fibers in tension wood in aspen stem *(Populus tremula L. x tremuloides Michx.)*. The slight plasmolysis, forming Hechtian strands similar with the structure of the "real" plasmatubules.

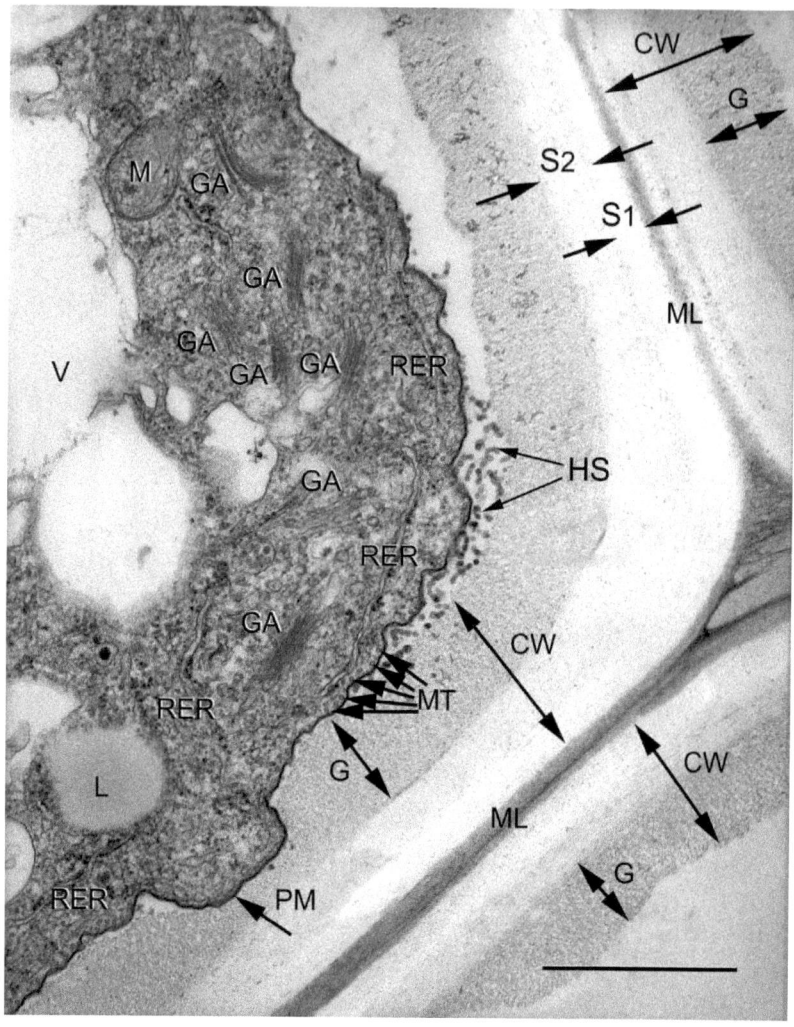

Fig. 11. The same in detail at high magnification. Activity organelles in forming a gelatinous layer of the cell wall: the Golgi apparatus, endoplasmic reticulum, plasmalemma - everything is in a highly excited state.

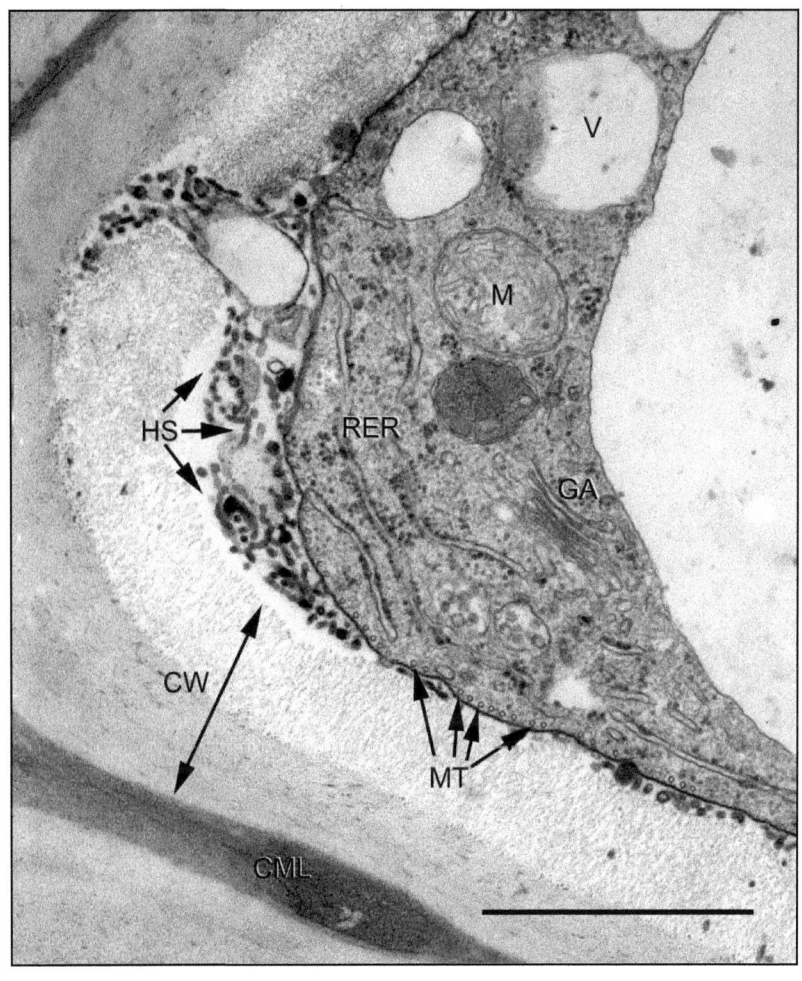

Fig. 12. The same in detail at high magnification. Activity organelles in forming a gelatinous layer of the cell wall.

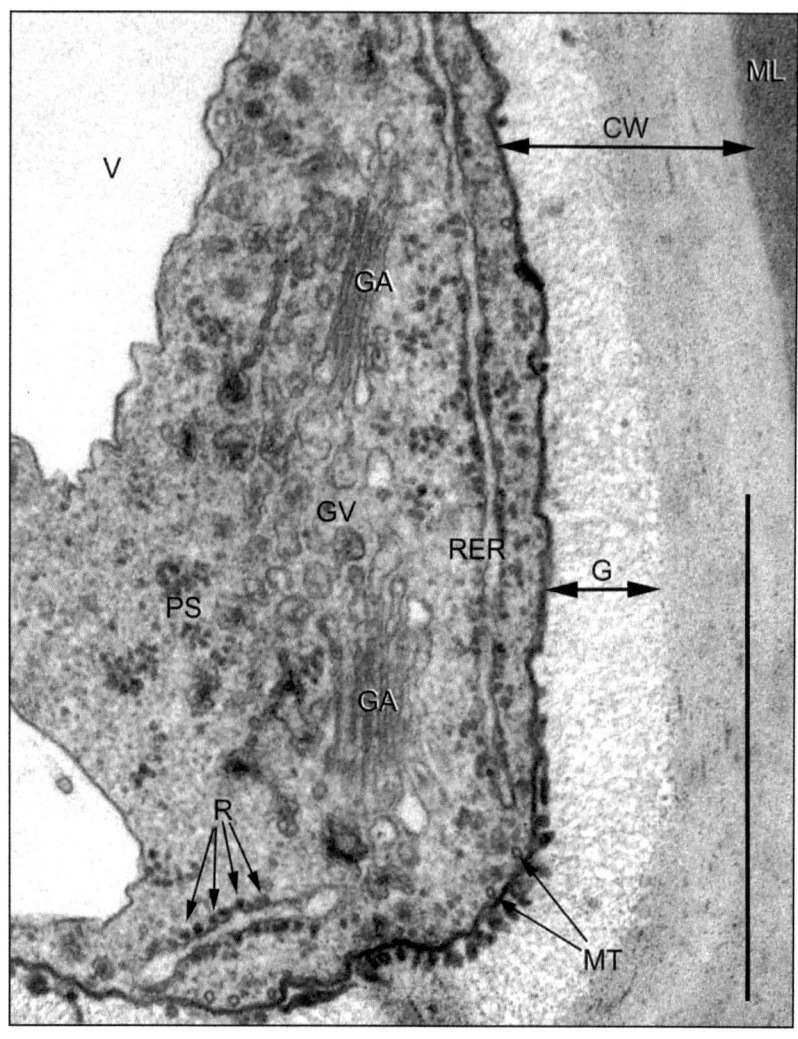

Fig. 13. The same in detail at high magnification. Activity organelles in forming a gelatinous layer of the cell wall. Hechtian strands visible destroyed in the initial stages of their formation.

.

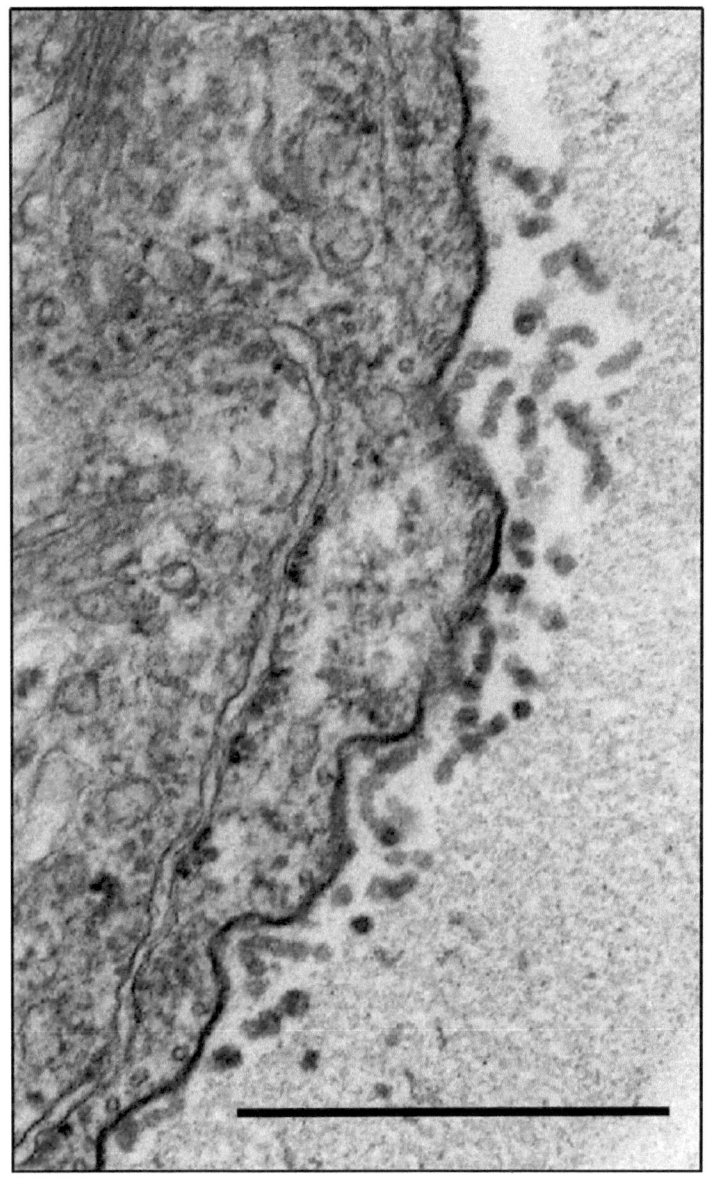

Fig. 13 a The same in detail at high magnification.

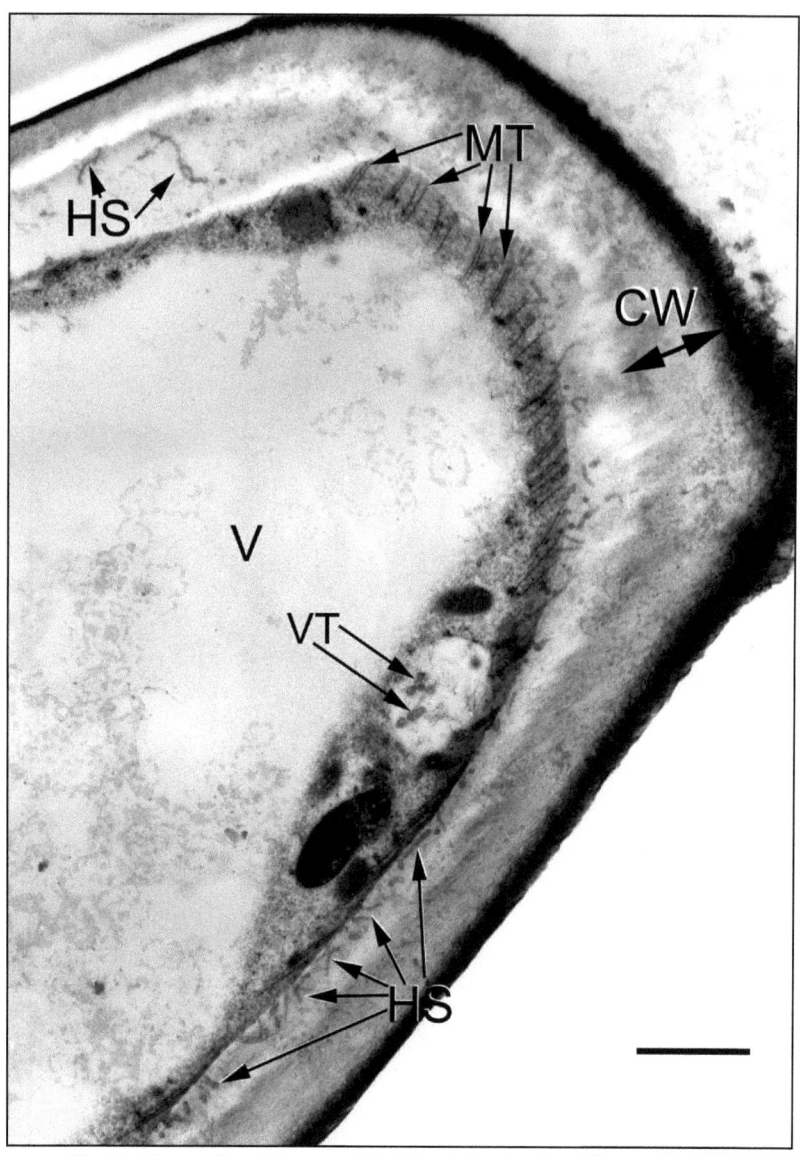

Fig. 14. The section (at an angle) of the hair of cotton (*Gossypium hirsutum* L.) seed (trichome).

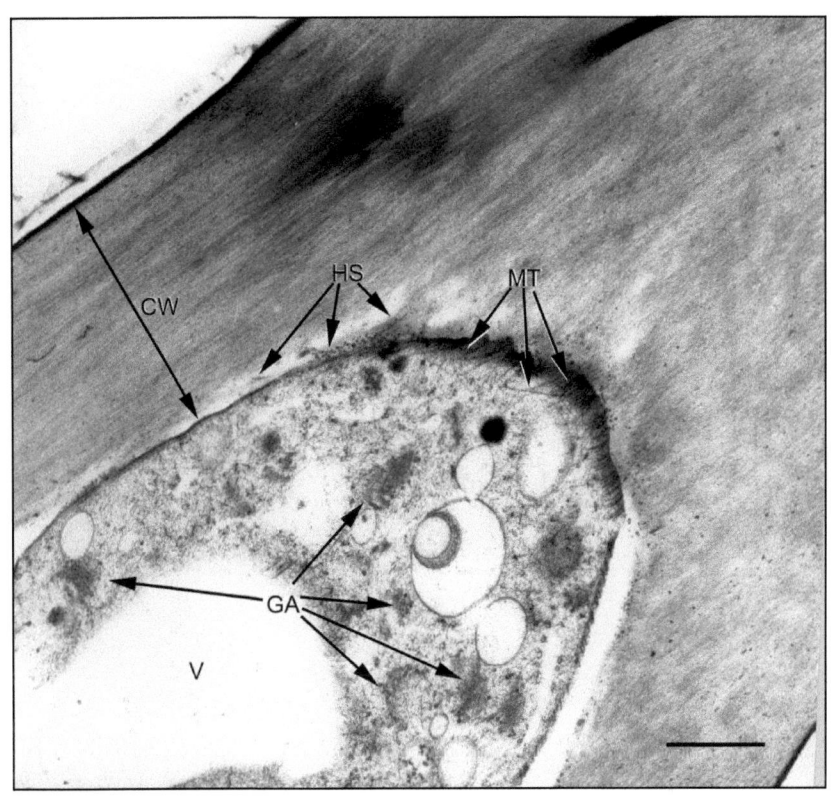

Fig. 15. The same. Numerous structures of the Golgi apparatus,
microtubules and Hechtian strands.

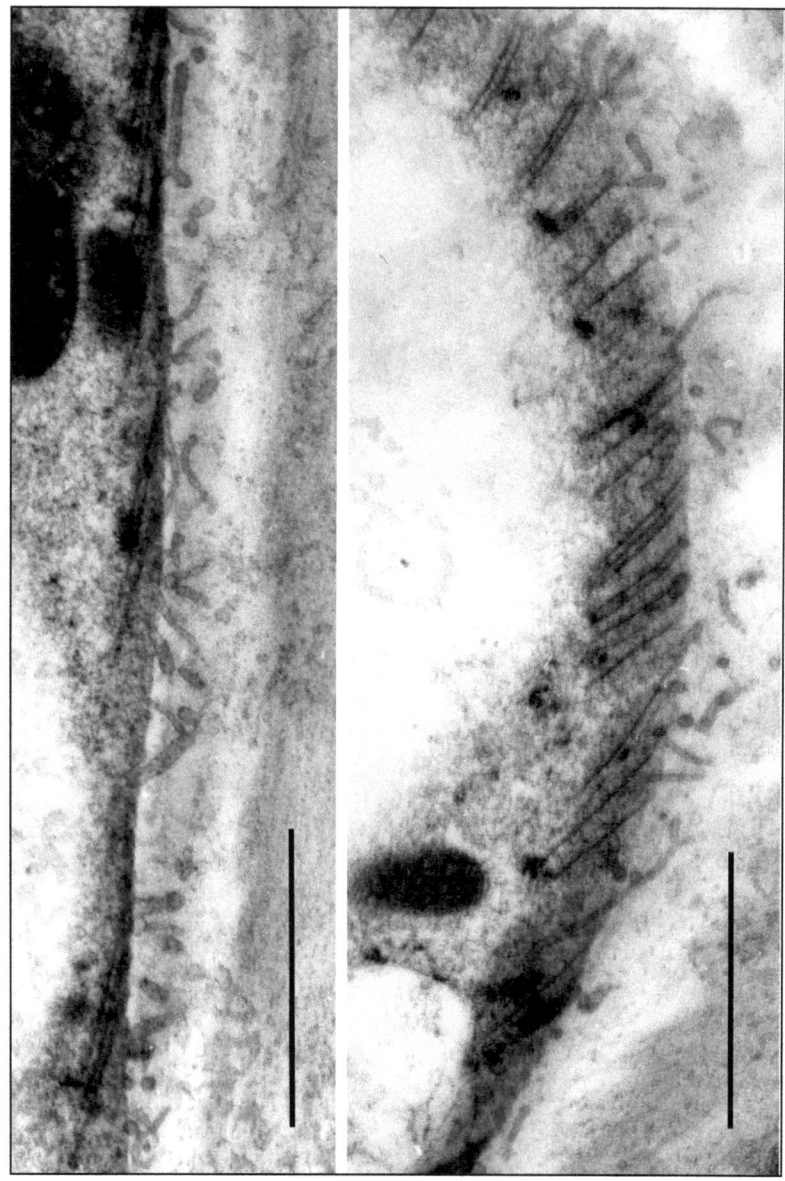

Fig. 16. The same in detail at high magnification. Hechtian strands scattered between the plasma membrane and cell wall.

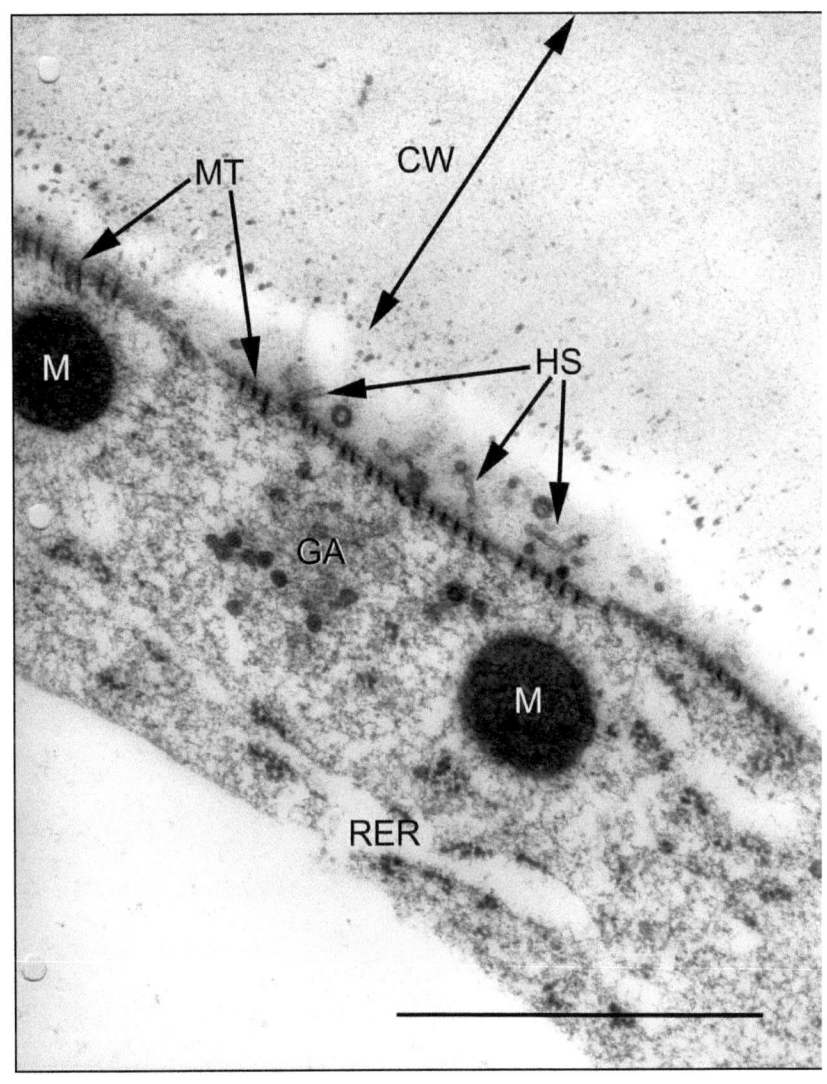

Fig. 17. The same in detail at high magnification.

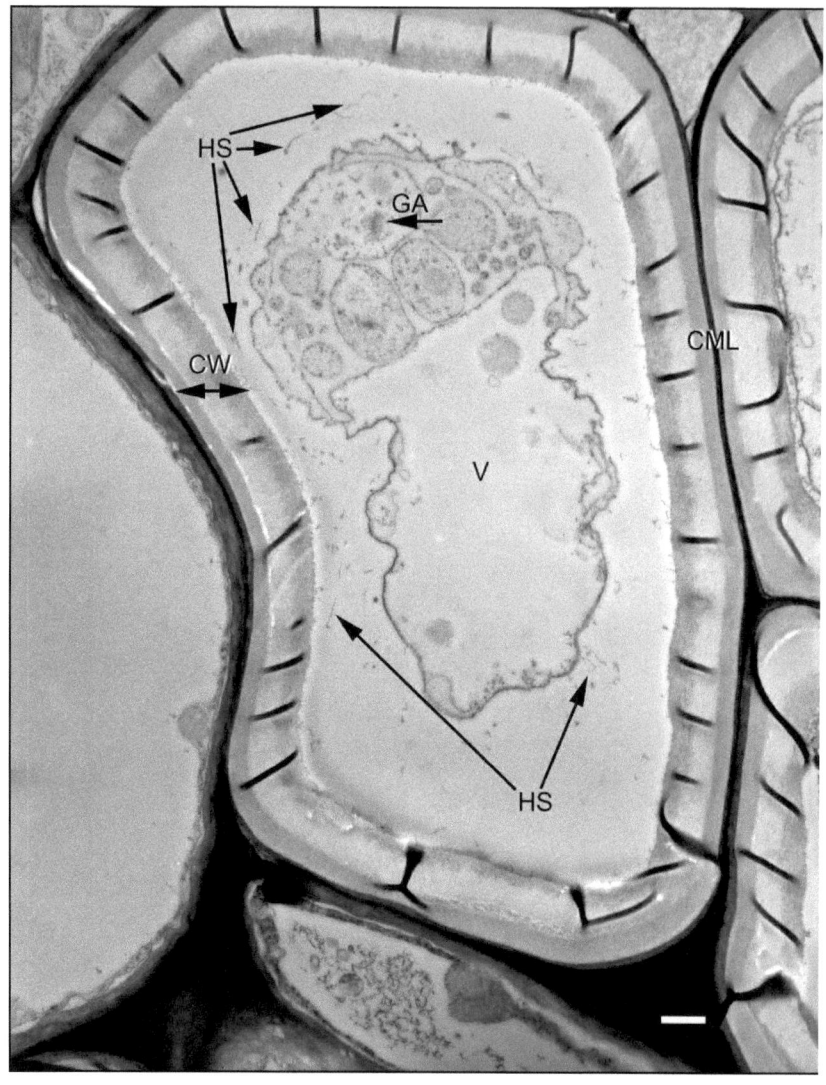

Fig. 18. Cross section of secondary phloem fibers of hemp (*Cannabis*). There is a strong plasmolysis and fragments of the Hechtian strands.

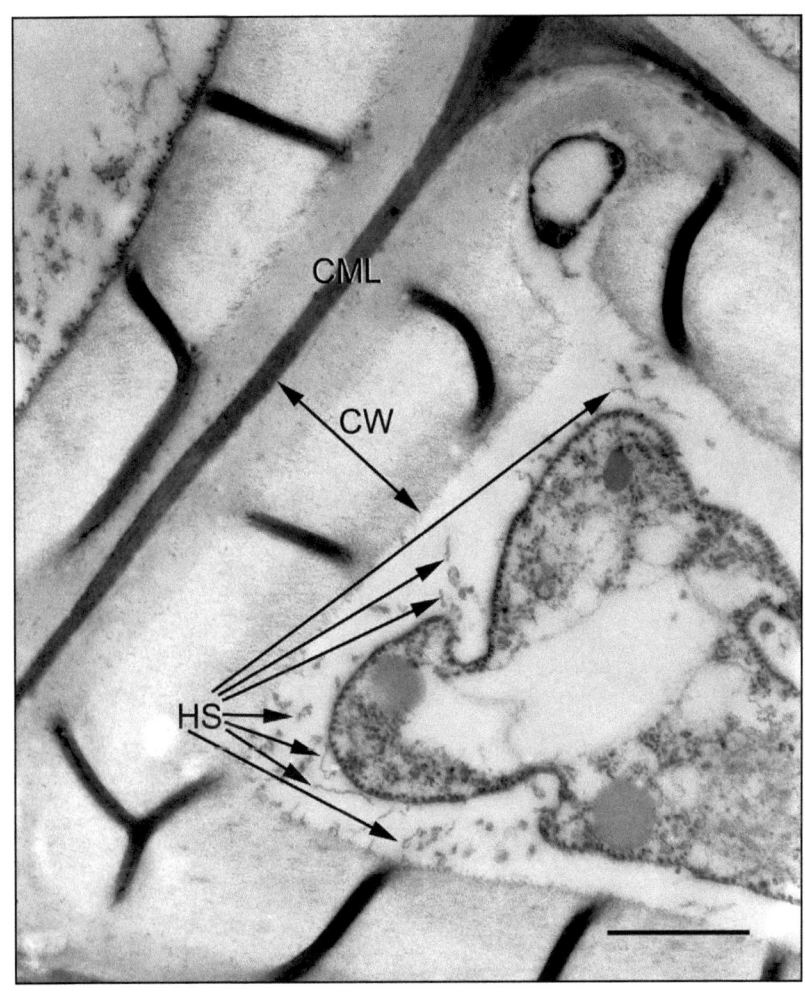

Fig. 19. The same in detail at high magnification.

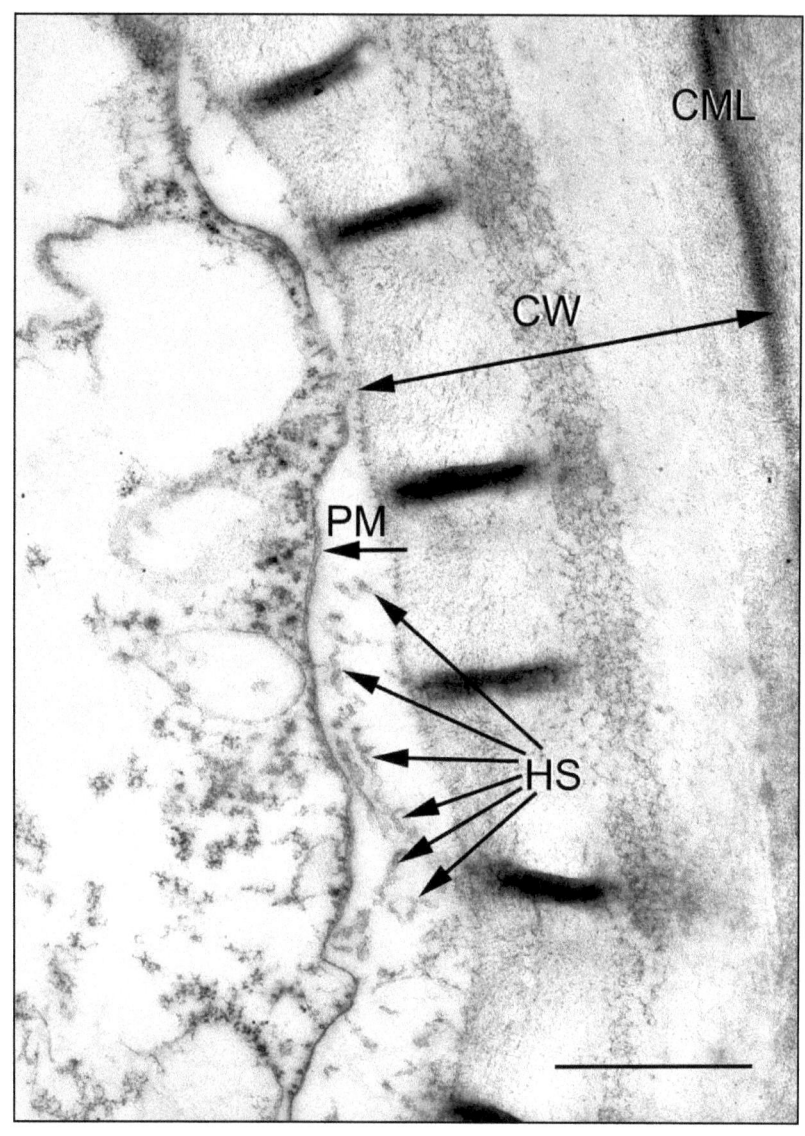

Fig. 20. The same in detail at high magnification.

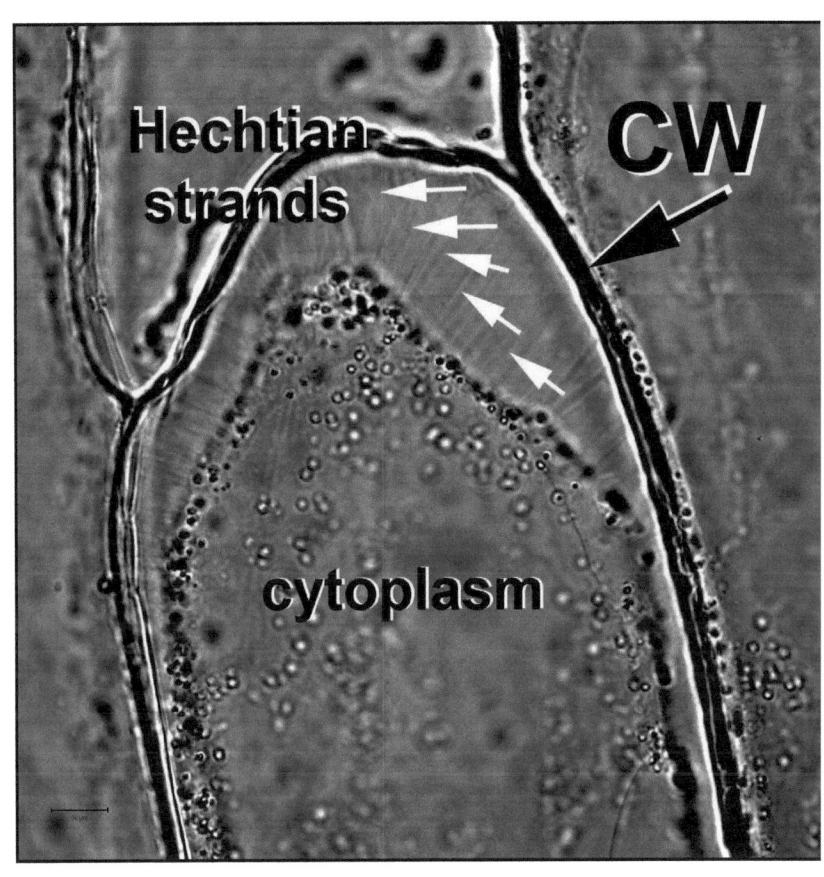

Fig. 21. Plasmolysis in Pea cells.

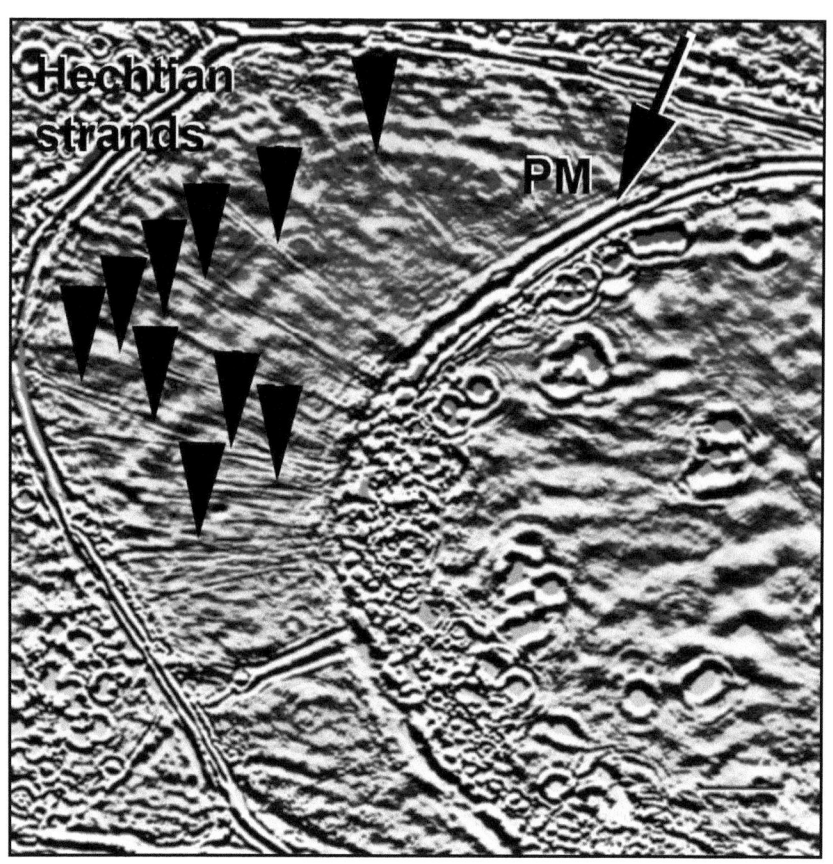

Fig. 22. The same.

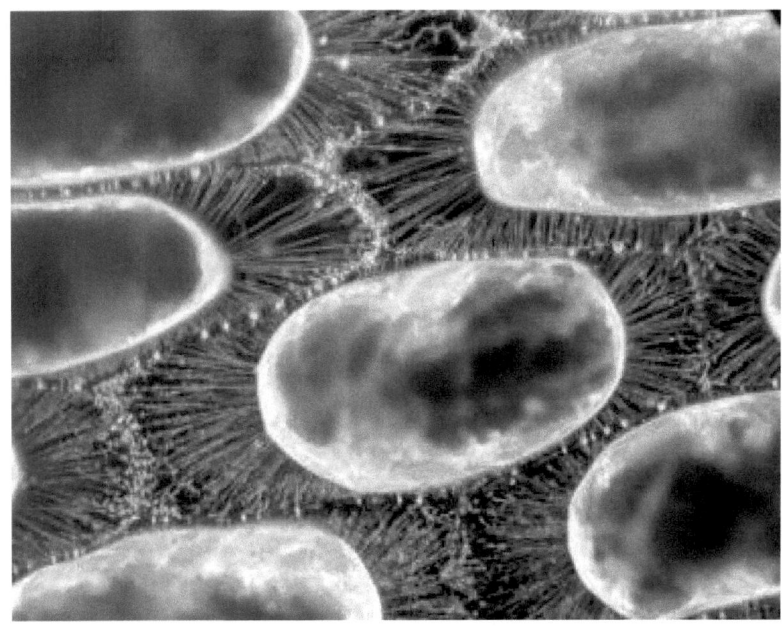

Fig. 23. Plasmolysis in onion cells. (From Onion scale leaf epidermal cells plasmolysed in 0·75 m mannitol. Stretched strands of plasma membrane connect the protoplast to the cell wall; small sections of the plasma membrane and peripheral cytoplasm remain tethered to the wall. Confocal projection, × 750. Photograph courtesy of Brian Gunning. Plant, Cell & Environment Volume 25, Issue 2, pages 239-250, 11 MAR 2002 DOI: 10.1046/j.0016-8025.2001.00808.x).

http://onlinelibrary.wiley.com/doi/10.1046/j.0016-8025.2001.00808.x/full#f5

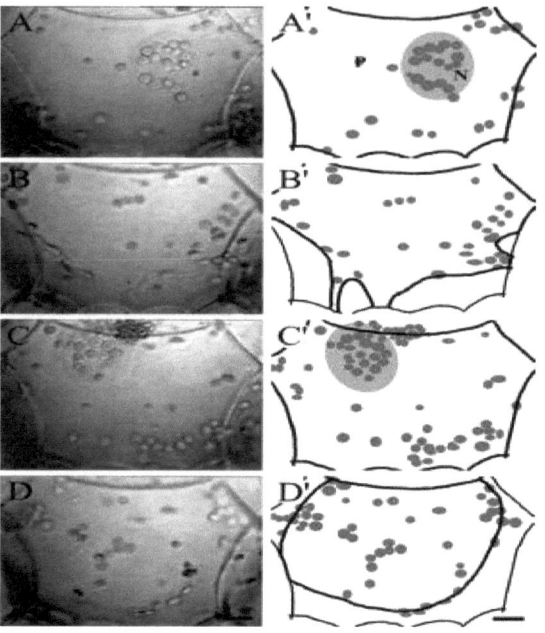

Fig. 24. Concave- and convex-type plasmolysis induced by hypertonic treatment on the end wall of a mesophyll cell. Optical images of the cytoplasmic layer along the end wall were captured before hypertonic treatment (A), at the time of concave-type plasmolysis induced in APW containing 0.5 M sorbitol (B), immediately after deplasmolysis in fresh APW (C), and at the time of convex-type plasmolysis induced by the second hypertonic treatment with APW containing 0.6 M sorbitol (D). The outlines of cell wall and plasma membrane were traced and are illustrated in A'–D'. N, nucleus; P, plastid. Bars = 10 μm   (Hayashi e. a., 2003, 2006).

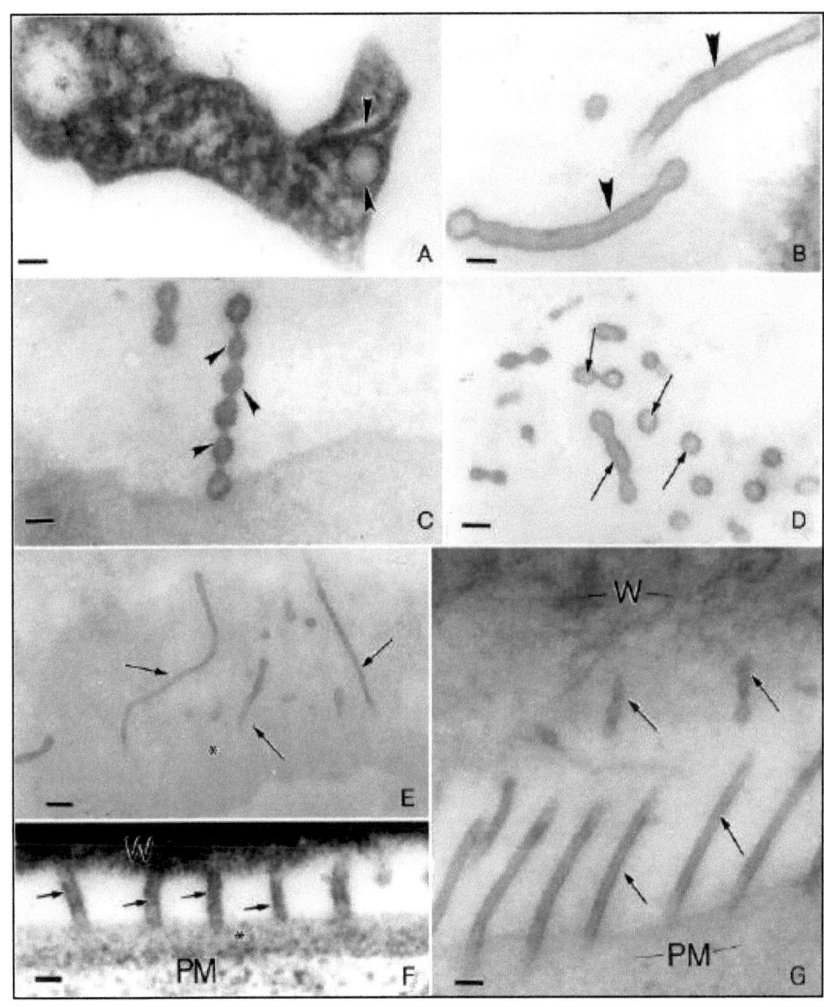

Fig. 25.

63

Fig. 25. Hechtian strand ultrastructure. (A) Longitudinal section of a Hechtian strand after 1 min of plasmolysis. The strand includes distinct cytoplasm and what appears to be endoplasmic reticulum (arrowheads). Scale bar, 48 nm. (B) Longitudinal section of Hechtian strand after 10 min of plasmolysis. Note that the strands (arrowheads) are very thin and devoid of any noticeable organelles or cytoskeletal structures. Scale bar, 50 nm. (C) Longitudinal section of Hechtian strand after 30 min of plasmolysis. The strand (arrowheads) begins to transform (i.e. degenerate) into a linear array of "beads." After longer periods of plasmolysis, very few strands can be found. Scale bar, 46 nm. (D) Cross-section of Hechtian strands after 30 min of plasmolysis. Note that the strands (arrows) are devoid of organelles and cytoskeletal components. Scale bar, 77 nm. (E) PTA staining of Hechtian strands. The distinctly stained strands (arrows) can be seen emerging from the unstained PM (*). Scale bar, 130 nm. (F) Junction zone of Hechtian strands from PM to CW after 2 min of plasmolysis. Note the strands (arrows) emerging from the plasma membrane (PM) and attaching to the CW (W). No distinct structures can be seen at the interface of the strand with the CW. Scale bar, 55 nm. (G) PTA-stained Hechtian strands in cell plasmolyzed for 5 min. Note the strands (arrows) emerging from the PM and attached to the inner fibrous zone of the CW (W). Once again, no distinct connections can be seen between the strand and CW. Scale bar, 50 nm.

(From Domozych, D.S., Roberts, R., Danyow, C., Flitter, R., Smith, B. and Providence K. (2003) Plasmolysis, Hechtian strand formation and localized membrane-wall adhesions in the desmid, Closterium acerosum (Chlorophyta). Journal of Phycology. 39:1-18.).

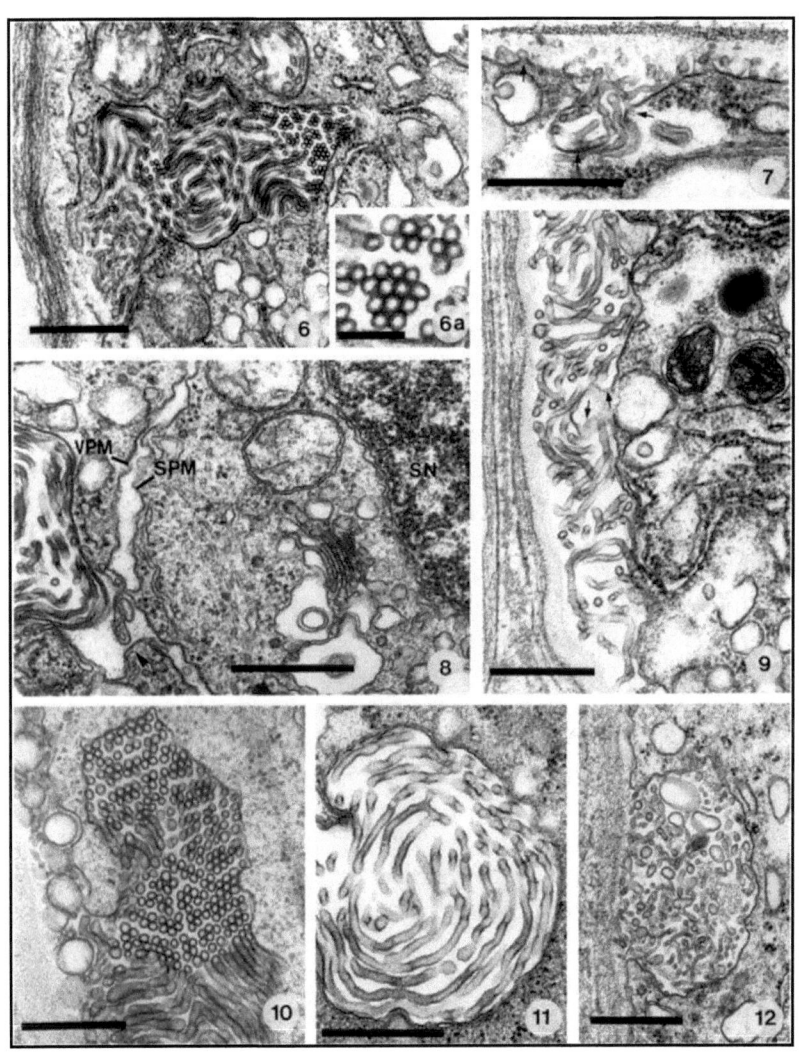

Рис. 26 (Рис. 6-12. According to a published article of Kandasamy e.a., 1988).

.

Electron micrographs of different portions of pollen tubes. Bars = 0.5 μm; **Fig. 6a**, bar = 0.1 μm. **Fig. 6.** Semivivo-grown pollen tube. Convoluted plasmatubules enclosed by a pocket-like plasma-membrane invagination. **Fig. 6a.** Portion of **Fig. 6** enlarged. Note the circular profiles of the tubules and their arrangement in small groups containing varying numbers of tubules in the cross-sectional view. **Fig. 6,** × 59000; **Fig. 6a,** × 148000. **Fig. 7.** Semivivo-grown pollen tube. *Arrows* indicate the continuity of plasmatubules with the plasma membrane. × 59000. **Fig. 8.** Semivivo-grown pollen tube. The plasma membrane ensheathing the pocket-like invagination appears to be continuous with the outer membrane of a sperm cell (*arrow*). *VPM*, Vegetative cell plasma membrane; *SPM*, sperm cell plasma membrane; *SN*, sperm nucleus. × 54625. **Fig. 9.** Cross section of an in-vivo-grown pollen tube (24 h). Convoluted plasmatubules in the periplasmic space between the pollen-tube wall and plasma membrane. Note the clear evagination of the plasma membrane in the region where a vesicle appears to fuse with it (*arrows*). × 45670. **Fig. 10.** In-vivo-grown pollen tube (24 h). Pocket-like invagination filled with closely packed plasmatubules (longitudinal and cross-sectional profiles). × 46140. **Fig. 11.** In-vivo-grown pollen tube (4 h). Pocket-like invagination containing occasionally branched plasmatubules. × 54500. **Fig. 12.** In-vivo-grown pollen tube (24 h). Plasma membrane invagination containing vesicular and tubular structures. × 39300

Planta (1988) 173:35–41. Planta. 9 Springer-Verlag 1988.
Plasmatubules in the pollen tubes of Nicotiana sylvestris
. M.K. Kandasamy, R. Kappler and U. Kristen*.
http://link.springer.com/article/10.1007%2FBF00394484#page-1

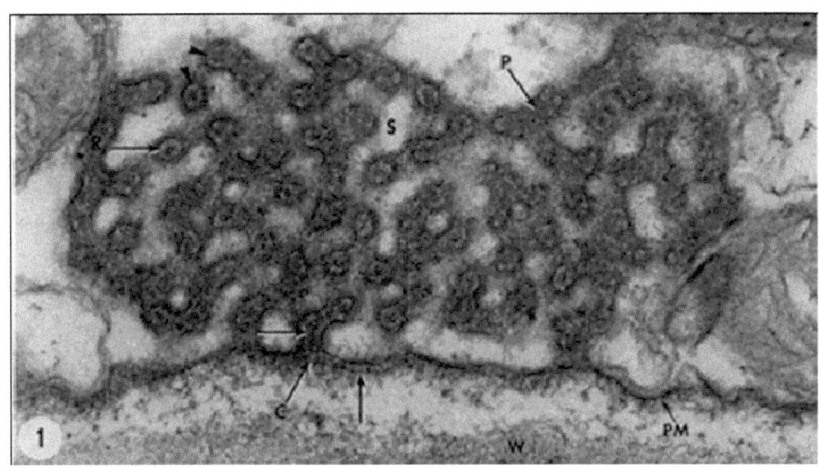

Fig. 27. Charosome structure. (From Pesacreta T.C., Lucas W.J. Plasma membrane coat and a coated vesicle-associated reticulum of membranes: their structure and possible interrela-tionship in Chara corallina. J. Cell. Biol. 1984 Apr; 98(4): 1537-45.).

http://www.ncbi.nlm.nih.gov/pubmed/?term=Thomas+C.+Pesacr eta+and+William+J.+Lucas+plasma+membrane+Coat

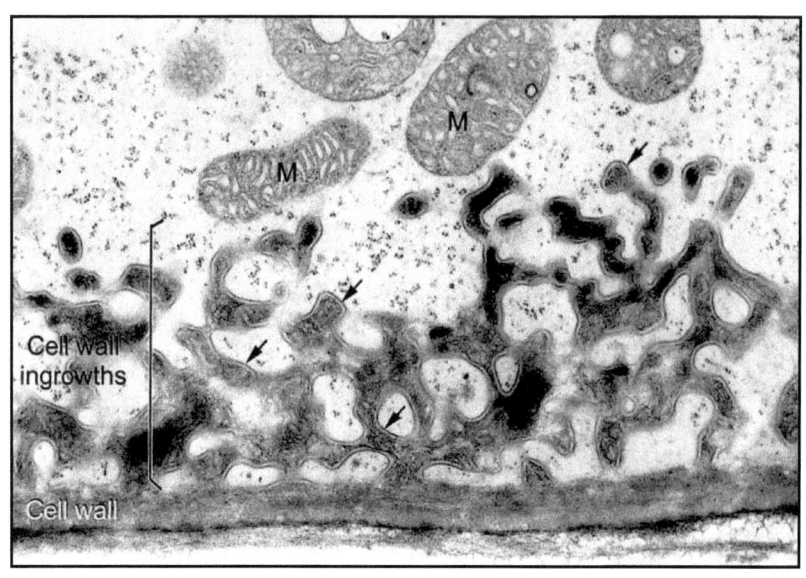

Fig. 28. Transfer cell wall. Elaboration of a cell wall into projections that are lined by plasma membrane (arrows), thus providing an enhanced surface area for exchange of many different types of solutes. Transfer cells develop in various plant tissues involved in transport. Mitochondria (M) are usually found in the vicinity of the wall labyrinths. (Figure 7 from A. Browning and B.E.S. Gunning, freeze-substituted transfer cell in the haustorium of a *Funaria* sporophyte, based on Gunning BES, Steer MW (1996) Plant Cell Biology: Structure and Function. Jones and Bartlett, Boston. pp 1-131. (German translation, Biologie der Pflanzenzelle, Gustav Fischer Verlag, Stuttgart, 1995).).

http://plantsinaction.science.uq.edu.au/edition1//?q=figure_view/196

Paramural bodies and other entities of the plasma membrane at the formation of the secondary cell walls of plants.

Salnikov V.V.

1. Kazan (Volga region) Federal University). 420008, Kremlevskata str., 18, Kazan, Russia.

2. Lobachevsky str., 2/31, Kazan, Russia, 420111. Kazan Institute of Biochemistry and Biophysics of the Kazan Scientific Center of the Russian Academy of Sciences. Tel: +7(843) 2 31 90 34; fax: +7(843)2 92 73 47.

e-mail: salnikov_russ@yahoo.com

Structures with the membrane and localized between the plasma membrane and cell wall, are named paramural bodies. A comprehensive analysis of the ultrastructure of paramural bodies and other entities in the plasma membrane at the formation of the secondary cell walls of plants is made. The objectives of our research on the formation of paramural bodies were: to analyze the ultrastructure of the formations found between the plasma membrane and the secondary cell wall during the formation of the polysaccharide cell wall of flax phloem fibers, tension wood of poplar, seed hairs of cotton (trichome), secondary phloem fibers of hemp), to determine the origin (the cause) of these paramural bodies. The above objects are the classic examples, by which we study the formation of the secondary cell walls. A long standing material that we obtained studying of the ultrastructure of the process on these sites allows us to fully analyze the observed structure formation. The accumulated in our research material on studying the formation of the secondary cell walls of plants of different species has allowed us to contribute to the debate about some causes of plasmatubules origin in the area between the plasma membrane and cell wall, dynamics of their formation and structure. In the cases under consideration at the formation of the secondary cell wall, we deal to a greater extent with the structures arisen as a result of plasmolysis. The author concludes that the structures previously taken for plasmatubules in most cases are the Hecht's threads.

Acknowledgements:

Dr. T.A. Gorshkova - Professor of Kazan Institute of Biochemistry and Biophysics of the Kazan Scientific Center of the Russian Academy of Sciences, Russian Federation.

Dr. Candace Haigler - Professor of Crop Science and Plant Biology Department of Crop Science North Carolina State University, USA.

Dr. Mark Grimson - Department of Biological Science Texas Tech University, Lubbock, USA.

Dr. Ewa J. Mellerowicz – Professor Department of Forest Genetics and Plant Physiology Swedish University of Agricultural Sciences (Sveriges lantbruksuniversitet) S901-83 Umea, Sweden

Dr. Ageeva M.V. - Kazan Institute of Biochemistry and Biophysics of the Kazan Scientific Center of the Russian Academy of Sciences, Kazan, Russian Federation.

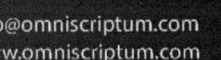 

Printed by Books on Demand GmbH, Norderstedt / Germany